Multilingual Information Management

Information, Technology and Translators

CHANDOS
INFORMATION PROFESSIONAL SERIES

Series Editor: Ruth Rikowski
(email: Rikowskigr@aol.com)

Chandos' new series of books is aimed at the busy information professional. They have been specially commissioned to provide the reader with an authoritative view of current thinking. They are designed to provide easy-to-read and (most importantly) practical coverage of topics that are of interest to librarians and other information professionals. If you would like a full listing of current and forthcoming titles, please visit www.chandospublishing.com.

New authors: we are always pleased to receive ideas for new titles; if you would like to write a book for Chandos, please contact Dr Glyn Jones on g.jones.2@elsevier.com or telephone +44 (0) 1865 843000.

Multilingual Information Management

Information, Technology and Translators

Ximo Granell

AMSTERDAM • BOSTON • HEIDELBERG • LONDON
NEW YORK • OXFORD • PARIS • SAN DIEGO
SAN FRANCISCO • SINGAPORE • SYDNEY • TOKYO
Chandos Publishing is an imprint of Elsevier

Chandos Publishing is an imprint of Elsevier
225 Wyman Street, Waltham, MA 02451, USA
Langford Lane, Kidlington, OX5 1GB, UK

Copyright © 2015 Elsevier Ltd. All rights reserved.

No part of this publication may be reproduced or transmitted in any form or by any means, electronic or mechanical, including photocopying, recording, or any information storage and retrieval system, without permission in writing from the publisher. Details on how to seek permission, further information about the Publisher's permissions policies and our arrangements with organizations such as the Copyright Clearance Center and the Copyright Licensing Agency, can be found at our website: www.elsevier.com/permissions.

This book and the individual contributions contained in it are protected under copyright by the Publisher (other than as may be noted herein).

Notices
Knowledge and best practice in this field are constantly changing. As new research and experience broaden our understanding, changes in research methods, professional practices, or medical treatment may become necessary.

Practitioners and researchers must always rely on their own experience and knowledge in evaluating and using any information, methods, compounds, or experiments described herein. In using such information or methods they should be mindful of their own safety and the safety of others, including parties for whom they have a professional responsibility.

To the fullest extent of the law, neither the Publisher nor the authors, contributors, or editors, assume any liability for any injury and/or damage to persons or property as a matter of products liability, negligence or otherwise, or from any use or operation of any methods, products, instructions, or ideas contained in the material herein.

British Library Cataloguing-in-Publication Data
A catalogue record for this book is available from the British Library

Library of Congress Control Number: 2014955042

ISBN: 978-1-84334-771-2

For information on all Chandos publications
visit our website at http://store.elsevier.com/

Typeset by Thomson Digital

Printed and bound in the United Kingdom

Dedication

To Noelia Judit, Marina Odet
and Débora Alba

Contents

List of figures and tables	ix
Biography	xi
Foreword: Outside in the ICT machine	xiii
Acknowledgments	xv
List of abbreviations	xvii

1 Introduction   1

Part One   Multilingual information and ICT needs   3

2 Information and translators   5
     2.1 Information, communication and information systems   6
     2.2 Multilingual information professionals   16

3 Technology and translation   21
     3.1 Tools to support translators   23
     3.2 Translation tools: origins and evolution   24
     3.3 The translator's workstation   27
     3.4 CAT tools and freelance translators   32

4 Information Literacy and Multilingual Information Management   35
     4.1 Information Literacy and Multilingual Information Professionals   35
     4.2 Information Literacy defined   36
     4.3 Information Literacy models and perspectives   38
     4.4 Information Literacy in the workplace   39
     4.5 Training information literate MIPs   41

5 A strategic approach to adopt ICT: from using information and communication technology to making use of information and technology to communicate   45
     5.1 The Information Systems approach to ICT   45
     5.2 Information Systems strategy   47
     5.3 IS and ICT adoption in small businesses   49

Part Two   Multilingual information and perspectives on ICT   61

6 A research framework for Multilingual Information Management   63
     6.1 Informant domains   63
     6.2 Development of the conceptual framework   65
     6.3 A research model of Multilingual Information Management   69

7	**Research methods for studying multilingual information management: an empirical investigation**	**89**
	7.1 Research approaches	89
	7.2 Selecting a suitable approach	91
	7.3 How to explore ICT adoption and use	94
	7.4 How to analyse organisational impacts and evaluate ICT sophistication	98
	7.5 The data analysis scheme	105
8	**Needs and perspectives of multilingual information professionals: findings of an empirical study**	**115**
	8.1 Characteristics of MIPs	115
	8.2 ICT adoption	117
	8.3 CAT tool adoption	124
	8.4 The characteristics of freelance translators adopting CAT tools	131
	8.5 The characteristics of the freelance translation businesses adopting CAT tools	131
	8.6 Perceptions of ICT and perceptions of CAT Tools	134
	8.7 Impacts of CAT tool adoption	151
	8.8 Summary of needs and perspectives	153

Part Three   Multilingual Information Management: matching needs and perspectives   155

9	**From PLEs to PLWEs: a Multilingual Information Management System**	**157**
	9.1 Personal Learning Environments (PLEs)	157
	9.2 Personal Learning and Working Environments (PLWEs)	160
	9.3 A Multilingual Information Management System	161
	9.4 Structure of an MIMS	163

References	**165**
Appendix 1 Translators in the 21st century: a study of skills, software and strategies	183
Appendix 2 Online survey for CAT tools adopters	191
Appendix 3 Online survey for CAT tools non-adopters	203
Appendix 4 Addressing non-response bias: Mann-Whitney test between early and late respondents	213
Appendix 5 Qualitative analysis form	215
Appendix 6 Summary of qualitative data analysis	223
Index	**225**

List of figures and tables

Figures

Figure 2.1	Example of data, information and knowledge in a multilingual communication context	7
Figure 2.2	Shannon-Weaver's Model of Communication	9
Figure 2.3	Nida and Taber's Model of Communication and Translation	9
Figure 2.4	O'Hagan and Ashworth's Model of TMC	10
Figure 2.5	Jakobson's Theory of Communication	10
Figure 2.6	Organisational boundary-spanning information systems (Jessup and Valacich, 2005, p. 219)	15
Figure 3.1	Melby's translator's workstation (based on Melby, 1982)	28
Figure 3.2	Hutchins and Somers' dimensions of translation automation (Hutchins and Somers, 1992, p. 148)	28
Figure 3.3	Melby's eight types of translation technology (Melby, 1998)	29
Figure 3.4	IAMT Certification initiative classification (based on Hutchins, 2000b)	29
Figure 3.5	*Compendium of translation software* classification of translation tools (based on Hutchins, 2000a)	30
Figure 3.6	Austermühl's process oriented approach (based on Austermühl, 2001)	31
Figure 6.1	Informant domains	65
Figure 6.2	Overview of levels of ICT adoption in Multilingual Information Management	67
Figure 6.3	Overview of the conceptual framework	69
Figure 6.4	Research model: ICT adoption by freelance translators	70
Figure 6.5	Multilingual Information Management activities	71
Figure 6.6	Multilingual Information Management activities	85
Figure 7.1	IT strategy bipolar items	97
Figure 7.2	Data analysis for an exploratory study	106
Figure 7.3	Components of Miles and Huberman data analysis method (1994, p. 12)	111
Figure 7.4	Qualitative data analysis plan	111
Figure 8.1	ICT usage and familiarity and experience with ICT	124
Figure 8.2	Scree test for factor analysis of perceptions of ICT	136
Figure 8.3	Scree test	139
Figure 8.4	Descriptive matrix of the factors affecting CAT tool adoption among adopters	146
Figure 8.5	Descriptive matrix of the factors affecting CAT tool adoption among non-adopters	147
Figure 8.6	Predictor-outcome matrix of relevance of the construct "Compatibility" (COMP) for CAT tool adoption	149
Figure 8.7	Research model enhanced by empirical data	153
Figure 9.1	Conception of PLE (adapted from Castañeda and Adell 2013, author's translation)	158
Figure 9.2	Components of a PLE (adapted from Castañeda and Adell 2013, author's translation)	159
Figure 9.3	Content curation process	160
Figure 9.4	Components of an MIMS	163

Tables

Table 5.1	Motivators affecting IS adoption in SMEs	54
Table 5.2	Types of CEO involvement in computerisation in SMEs (Martin, 1989, p. 192)	54
Table 5.3	Inhibitors affecting IS adoption in SMEs	55
Table 5.4	Classes of factors affecting IS success in SMEs	58
Table 6.1	Document production activity	73
Table 6.2	Information search and retrieval activity	75
Table 6.3	Translation creation activity	77
Table 6.4	Communication activity	81
Table 6.5	Marketing and work procurement activity	82
Table 6.6	Business management activity	83
Table 6.7	Determinants of ICT adoption	86
Table 6.8	Impacts of specialised ICT adoption (Khandwalla, 1977)	86
Table 7.1	Items for Question 1: using terminology management tools/translation memory	103
Table 7.2	Items for Question 2: terminology management tools/translation memory and the translation sector	103
Table 7.3	Items for Question 3: learning about terminology management tools/translation memory	104
Table 7.4	Items for impacts of terminology management tools/translation memory	104
Table 8.1	ICT usage	118
Table 8.2	Familiarity and experience with ICT	122
Table 8.3	CAT tool vs. ICT adoption prediction (logistic regression model)	128
Table 8.4	CAT users and use of other ICT	129
Table 8.5	CAT users and familiarity with other ICT	130
Table 8.6	CAT tool adopters' characteristics (logistic regression model)	133
Table 8.7	Factor analysis results for perceptions of ICT	137
Table 8.8	Factor analysis results for perceptions of CAT tools	140
Table 8.9	A comparison of ICT and CAT tool factors	142
Table 8.10	CAT perceptions of adopters and non-adopters	143
Table 8.11	Perceptions of respondents with different levels of experience with CAT tools	145
Table 8.12	Perceived relative advantages associated with CAT tool adoption	150
Table 8.13	Positive impacts of adopting CAT tools (adopters)	151
Table 8.14	Impacts of adopting CAT tools (adopters and non-adopters)	152

Biography

Dr Ximo Granell is a lecturer and researcher in Information Science and Audiovisual Translation at Universitat Jaume I of Castellón, in Spain. Holding a PhD in Information Systems from Loughborough University, United Kingdom, and being a Translation and Interpreting graduate and a Business Management postgraduate, his academic background has been connected with the fields of information management, translation and information systems. Dr Granell has taught undergraduate and master's courses at several universities on information management, business information systems, translation technology and video game localisation, topics on which he has researched and published in international journals, books and conferences. He has also worked as a translator specialising in software, website and video game localisation, as well as in ICT, marketing and business communication translation. His research interests include developing information literacy in specialised translation contexts, applying information systems to the translators' community of practice, assessing information competences and evaluating the quality of scientific research. Dr Granell has participated as a researcher in a number of R&D funded projects about these issues, such as recent projects on assessing information competences of Social Science students in Higher Education or implementing information systems to improve translation and localisation processes in audiovisual translation.

Foreword: Outside in the ICT machine

In today's world, information is a powerful resource that is becoming increasingly important and uncontrollably abundant. Its exponential growth means that half the information that is currently available has been gathered in the last fifty years. But it is not only the amount and massive stockpiling of data that keeps our society in constant evolution, rather it is the rational, orderly, productive and intelligent use of information that constitutes the primordial factor driving development. Organisations, enterprises, social groups and, in a word, individuals need to develop the competence to transform all that mass of information into intelligent knowledge and productive know-how.

We have shifted swiftly from the Information Society to the Knowledge Society, in which the greatest capital lies in human beings' capacity to think and to create. Ideas, individual or collective initiatives and thoughtful and creative effort are of great value to our society. And this enormous potential is increased when it is shared – something that is made possible by the information and communication technologies, which enable human beings to become interconnected by networks so that they can combine their intelligence, knowledge and creativity.

Knowing how to select and give sense to information and use it in order to solve problems, handle new situations and continue learning is a key issue in the teaching and learning scenario and also the professional practice and development in contemporary society, for every one of us, worldwide.

Though the need is shared, since every community of practice generates, seeks, retrieves and uses the resources and sources related to the cognitive structure being researched or studied and the tasks being performed, it becomes necessary to undertake studies focused on real user communities.

This is the viewpoint in the present book, focused on translators, their multilingual information management needs and their use of Information and Communication Technologies (ICT), nowadays essential and unavoidable.

In this sphere of application, it is important to remember that translators are not only information users, but also information processors and producers. Translation is, above all, a specialised activity that demands a changing flow of information in diverse languages and on several topics. Translators are constantly faced with the challenge and the responsibility of becoming acquainted with and using the diverse means that now exist for the location, retrieval, handling and dissemination of information, and of using the extraordinary new resources that information and technologies have made available for their work. In other words, in today's rapidly changing environment, the translator faces new situations that have emerged from the exponential growth of technologies and the flood of documentary and information material that has arrived in their wake.

Ximo Granell's monograph is a step forward in this context, offering a precise study on ICT, and computer-aided translation (CAT) tools in particular. His reflections are based on an empirical study with freelance translators based in the UK, but the study is wisely linked with a literature review on Information Systems (IS) and Information Literacy (IL) paradigm, essential to contextualise and understand these professionals' working life environment.

The fruitful combination of theory and practice is an added value in this book. Because, though designed to meet the needs of a specific user group, namely translation professionals, the reflections can be useful in a broader context and for a number of information-related professional and training profiles. Indeed, remarkably, the author prefers to speak about Multilingual Information Professionals, and thus his view opens up broader venues.

All in all, Multilingual Information Management: Information, technology, and translators is a compelling contribution, a relevant interdisciplinary input to Translation Studies, ICT, IS Studies and Information Literacy Studies.

Dora Sales
Literary translator, Senior Lecturer in Information & Documentation Studies
Department of Translation and Communication
Universitat Jaume I (Spain)

Acknowledgments

I am indebted to all the people who have assisted and supported me in writing and publishing this book, one way or another. Along the way I have been able to further develop myself as a researcher, as a lecturer, as a translator and as an information professional, but above all, as a person. This personal development will be part of me throughout the rest my life and undoubtedly help me in my continuous learning process, which I expect to share with many others. Long live lifelong learning!

I would like to start thanking Dr Glyn Jones, George Knott and Harriet Clayton for their trust, time and help throughout the editorial process of publishing this book.

Special thanks go to Dora and Fede, who have always believed in me and from whom I have learned so much. Thank you for your priceless encouragement and kind friendship.

I am truly grateful to all my colleagues in academia from the Translation and Communication Department at Universitat Jaume I for their day-to-day support and advice. Special thanks in this sense go to those with whom I have shared lots of special moments and with whom I have learned so many things: Dr Dora Sales and Dr Enriqueta Planelles, from the Information Science area, and my fellow colleagues of the TRAMA research group: Prof Frederic Chaume, Dr Beatriz Cerezo, Dr Irene de Higes, Dr José Luis Martí, Dr Juanjo Martínez Sierra, Anna Marzà, Dr Ana Prats, Ana Tamayo and Glòria Torralba.

I cannot forget to express my gratitude to Prof Heather Fulford for her mentoring, advice and support during my early years in academia at Loughborough University; to my colleagues at the Business School such as Prof Malcom King, Prof John Wilson or Dr Dave Coates; as well as to my former colleagues and friends at Loughborough University, in particular Dr Arvind Yadav, Dr Matoula Papaioannou, Dr José Alberto Hernández, Dr Emmanuel Touloupis, Dr Christian Tuch, Dr Nacho Rendo, and Dr John Whitley.

I am particularly grateful to all the translators who shared their experiences with me at some point and provided their kind feedback as informants. Also to my students over these years, with whom I have shared many discussions and initiatives aimed at innovating in education. They have fed my hunger to learn more about didactics and keep developing myself as a lecturer.

Thank you to those institutions that have supported my research, such as the Engineering and Physical Science Research Council (EPSRC), Loughborough University, and Universitat Jaume I.

Finally, I want to thank my family and friends for their immense encouragement and faith in me. In particular, to my wife Deb for her unconditional support and endless patience; to my older daughter, Noelia Judit, for being the light and joy of my life that is always there for me; and last but not least, to my soon to be second daughter, Marina Odet, for adding even more happiness to our busy lives and who might even share a birthday with this book. All of you fill my life with inspiration day after day. Thank you!

Ximo Granell (Castellón de la Plana, 2014)

List of abbreviations

ACRL	Association of College and Research Libraries
ALA	American Library Association
ANOVA	analysis of variance
ANZIIL	Australian and New Zealand Information Literacy
CAT	Computer-aided Translation
CAUL	Council of Australian University Librarians
CEO	Chief Executive Officer
CMC	Computer-mediated Communication
CMS	Content Management Systems
COMP	Compatibility
DSS	Decision Support Systems
DTP	Desktop Publishing
EASU	Ease of Use
EBMT	Example-Based Machine Translation
EHEA	European Higher Education Area
EIS	Executive Information Systems
EMT	European Master's in Translation
FAHQT	Fully Automatic High Quality Translation
FAMT	Fully Automated Machine Translation
FTP	File Transfer Protocol
HAMT	Human-Aided Machine Translation
IAMT	International Association for Machine Translation
ICT	Information and Communication Technology
IL	Information Literacy
IM	Instant Messaging
InfoLiTrans	Information Literacy for Translators
IS	Information Systems
IT	Information Technology
JISC	Joint Information Systems Committee
LSP	Language Service Provider
MAHT	Machine-Aided Human Translation
MIM	Multilingual Information Management
MIMS	Multilingual Information Management System
MIP	Multilingual Information Professional
MS	Microsoft
MT	Machine Translation
OAS	Office Automation Systems
OPTIMALE	Optimising Professional Translator Training in a Multilingual Europe
P2P	Peer-to-peer
PLE	Personal Learning Environment
PLWE	Personal Learning and Working Environment
QA	Quality Assurance
RBMT	Rule-based Machine Translation

READ	Relative Advantage
REDE	Result Demonstrability
SISP	Strategic Information Systems Planning
SME	Small and Medium Enterprise
SMT	Statistical Machine Translation
TEnTs	Translation Environment Tools
TM	Translation Memory
TMC	Translation-mediated Communication
TPS	Transaction processing systems
TRIA	Trialability
URL	Uniform Resource Locator
VBA	Visual Basic for Applications
WYSIWYG	What You See Is What You Get

Introduction

This book explores a number of issues that have to do with the way that information and technologies are used within the community of practice of translators. These professionals are information facilitators among different languages and cultures and, broadly speaking, their work is in high demand in today's globalised world. Not only industry and market globalisation, but also other socio-economic changes, such as a closer collaboration between European countries, technological developments, the advent and consolidation of the Internet, the rise of electronic business, and the increase in the use of electronic documents have also contributed to the demand for translation services. Quality and time requirements are no strange words to translators, since they are required to produce high-quality translations in ever-shorter time periods. Running in parallel with the increasing demand for translation services, various organisational developments have had, and are indeed continuing to have, a considerable impact on the translation services sector. For example, many large organisations have been divesting themselves of in-house translation teams to focus on their core business to reduce costs, resulting in an increasing number of translation assignments being outsourced to freelance translation businesses.

As for any other information professionals, a wide range of information and communication technologies (ICT) are available to translators today, both in terms of general purpose tools and resources, and in terms of specialised aids that can be used by them. Computer-aided translation (CAT) tools are probably the clearest example of translator-specific computer tools designed to increase their productivity and efficiency, and thus helping them to meet the demand for their services. While there has been much discussion among translators and other stakeholders of the language market about how translators make use of available ICT and information resources, as well as about the suitability of using software like CAT tools, few studies have empirically investigated their use by this community of practice. Research has rather focused on how translators should organise their working environment, how to automate translation processes, and the analysis of the technical features of CAT tools, or their use in large translation departments.

This book draws on previous empirical research undertaken to investigate the uptake of ICT, and CAT tools in particular, by freelance translators based in the UK and their perceptions on the latter[1] to frame their use of information and technology within an Information Systems (IS) and an Information Literacy (IL) paradigm that allows an explanation of their working life environment. In other words, the aim of this work is to

---

[1] The investigation was part of the research project "The adoption of translation software by translation SMEs: a study of productivity and organisational issues" (GR/R71795/01), led by Dr. Heather Fulford at Loughborough University and funded by the Engineering and Physical Sciences Research Council (EPSRC) of the United Kingdom.

provide practitioners with a broad and strategic perspective of how information and technology can help them exploit their ICT resources to become more productive and competitive in today's market. It is envisaged that the information in this book might not only be of interest to experienced and newly qualified translators, translators' educators, professional bodies for translators, and developers of CAT tools, but also to a wider audience of information-related businesses and information literacy educators. Although the focus is placed on the practice side, theoretical and methodological contributions are also present and this book also aims to nourish existing research, both in the translation studies and information management fields, with a research framework and models about ICT adoption and usage and a validation of instruments to measure ICT adoption based on empirical evidence.

The book is divided into three interrelated parts. In the first one, the background and context information is provided through a literature review of translation processes, technology, information literacy and information systems issues; i.e. a state of the art is focused on information and technology needs among multilingual information professionals. The second part of the book is the core element of this work and addresses the topics under study by applying research methods to investigate the perceptions of the translators' community of practice; i.e. translators' perspectives towards their use of information and technology are discussed in relation to their working environment by drawing on the findings of the empirical research work undertaken, the research framework and methodology designed to this end, and an interdisciplinary review of the literature. Last but not least, the third part of the book is aimed at bridging the gap between the needs of multilingual information professionals and their perspectives of information and technology by leading the discussion towards a strategic proposal for applying an IS approach to their work, outlining implications for translators' practice and training.

Part One

Multilingual information and ICT needs

"Computers will never replace translators, but translators who use computers will replace translators who don't."

(Hunt, 2002, p. 49)

In reviewing literature related to translation, information systems and information literacy, a number of themes could be seen to characterise multilingual communication, each relating to ways in which information and knowledge are perceived. Part One presents the background from a number of perspectives that contribute to a holistic view of the ICT-based environment that surrounds multilingual information related workplaces, mostly of freelance translators, the information literacy paradigm of their professional practice, and a strategic approach that helps big and small organisations to analyse their working setting to incorporate the information and communication tools they might need.

Information and translators

> "There is nothing like looking, if you want to find something. You certainly usually find something, if you look, but it is not always quite the something you were after."
>
> **(J.R.R. Tolkien, The Hobbit)**

One of the main problems that translators face today is the vast amount of information available to them. Although J.R.R. Tolkien wrote the quotation above in the 1920s, and in the context of a fantasy novel, it could nevertheless be applied to many real situations today when a very specific piece of information is required. As highlighted by experts from the global community of information professionals, "a big gap [is observed] between the quantity of content we have and the quality of search results we get when we need something specific. The feeling of searching for a needle in a haystack got the part of our everyday life, with a haystack that gets bigger and bigger every day, every hour and every minute" (Molnar, 2014). This, in the case of translators, is a rather common circumstance given the amount of information they need to process as language facilitators (c.f. section 2.2). We live in a globalised and digital world with a great deal of linguistic and cultural diversity. Socio-economical changes and technological advances, among other factors, have multiplied the demand for multilingual communication. Consequently, the need for translation services has increased significantly, as has the amount of all sorts of information available.

As a result of this exponential increase in the quantity of information and the number of channels through which it is delivered, we have come up against an information overload. This term, started to be popularised by Alvin Toffler in his 1970 book *Future Shock* (Toffler, 1970) and also referred to as "infoxication" later on (Cornellá, 1999, p. 42), has deserved much attention from the business management community due to the potentially negative impact on productivity and decision-making.[1] Therefore, a critical appraisal of what is valid information and how to access it quickly is a basic requirement for translators (Sales, 2008, p. 43), since they are not only information users, but also information processors and producers (Pinto, 1999, p. 106). This means that in today's world of an ever-increasing flow of information translators are required to produce high-quality translations in ever-shorter time periods to cope with an ever-larger workload (Pavlovich, 1999, p. 37; Schäffner, 2000, p. 7). It seems rather obvious that translators must be able to search, access, select and use information effectively, something that is closely related to the concept of "information literacy,"

---

[1] A simple search of the term "information overload" in Google Books' titles draws 260 results and the same search in Google Academic's titles about 764 articles and books (as for 14/02/2014). For academic essays theorising about this concept see, for example, Lincoln, (2011).

discussed in more detail in chapter 4 of the book, and the need to develop their information competences. The more information literate a person is, the less infoxicated he or she will be (Cornellá, 1999).

2.1 Information, communication and information systems

In this section, the three keys elements of information, communication and information systems are presented and discussed to lay the foundations of a "Multilingual Information Management System" (MIMS), that is, an information system for information professionals working in multilingual environments.

2.1.1 Information

A common simple definition of information is "the result of processing data" (Davis, 2000, p. 71). Data are usually considered to be raw facts, observations and representations of events, people, resources or conditions without much value until they have been processed and transformed into information (Bocij et al., 1999, p. 5). Information involves adding meaning to provide understanding, insight, conclusion, decision, confirmation or recommendation (Davis, 2000, p. 71), and therefore is a complex term that can be understood and described from many perspectives (Buckland, 1991, p. 351; Case, 2012, p. 42). For instance, Buckland distinguishes three principal uses of information: as a process (i.e. becoming informed), as knowledge (i.e. the knowledge communicated concerning some particular fact, subject or event), and as a thing (i.e. objects such as data or documents that contain knowledge or communicate information).

If we look at the common use of this term, there are several definitions for information (Collins English Dictionary, 2014):

- knowledge acquired through experience or study;
- knowledge of specific and timely events or situations, news;
- the act of informing or the condition of being informed;
- the meaning given to data by the way in which they are interpreted.

In them, we find several meanings that match the uses highlighted by Buckland, and the last reads that information is "the meaning given to data by the way in which they are interpreted", i.e. how data are interpreted in a meaningful context. This point stresses the subjective nature of information, as the same data can be interpreted differently by different people or in different situations, thus the relevance of placing it within a context that determines its value. Similarly, the process followed to add value to data, i.e. the abovementioned interpretation, implies a treatment process, a reasoning that contextualises raw data and transforms them into valid knowledge to serve a specific purpose. Information, therefore, is produced in response to an information need to shape knowledge.

According to Davis (2000, p. 71), knowledge can be defined as "information organized and processed to convey understanding, experience, accumulated learning, and expertise." Following this definition, the author classifies it into procedural knowledge (how to do something), formal knowledge (general principles, concepts, and procedures), tacit knowledge (expertise from experience that is somewhat hidden), and meta knowledge (knowledge about where knowledge is to be found).

From a pragmatic point of view, the context being addressed in this book is that of multilingual communication, so it is important to understand these three concepts in relation to this scenario. Examples of data at a very basic level of abstraction would be the visual symbols of a language system that make up its alphabet (i.e. letters, numbers, symbols). Information would then come when these symbols are put together and are given a meaning established through linguistic, social or cultural conventions (i.e. words, signs). And knowledge would be generated when the words of a language are put together to shape an idea in someone's mind with a specific purpose (i.e. to understand the relationships between different pieces of information to express how to do something, to explain a phenomenon, or to describe how to access more knowledge). At a higher level of analysis, data would refer to all representations of phenomena in a particular language, whether they are in the form of text, numbers, codes, graphs or pictures. Information is then produced through the interpretation of such phenomena in the context determined by a specific culture, a particular domain/subject-matter, or the conventions that guide the practice of some professions (i.e. the materialisation of concepts through signs in the form of (mostly) linguistic terms used continually by a social group). Knowledge would be finally generated by using pieces of information to articulate complex ideas, whether tacitly or explicitly, arising from an expertise in something that is required to be communicated within a specific context (i.e. using specialised terminology, in addition to terms used in common communication, to convey ideas with a purpose which is defined by the communicative context).

Figure 2.1 shows a representation of these concepts from the latter level of analysis in a multilingual setting. In the medical field of expertise, raw data can be a certain amount of glucose (glucose level) expressed through a numeric value (160), a mathematical symbol that indicates "higher than" (>) and an arbitrary unit of measurement that weights the amount (mg/dl). This would read as "a glucose level higher than 160 milligrams per decilitre," which isolated from other data simply provides an indicator

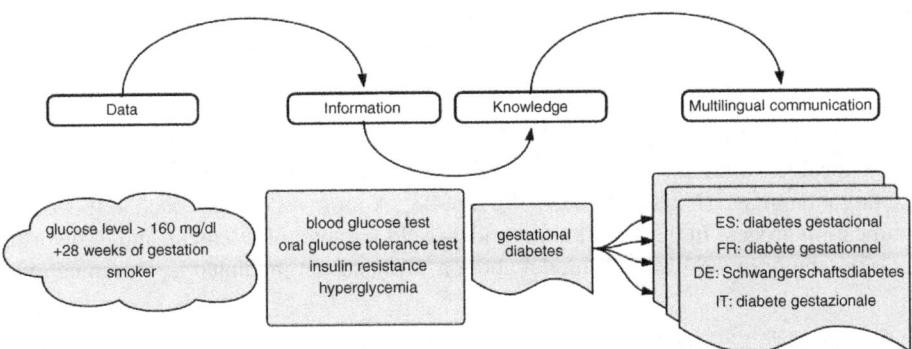

Figure 2.1 Example of data, information and knowledge in a multilingual communication context

of the amount of glucose (sugar) present in the blood of a human. That information combined with other data, such as being pregnant after 28 weeks of gestation and being a smoker, to the eyes of a midwife or a doctor, provide symptoms of a potential disease. Pieces of information start to be formed when such data are placed in the context of a screening test (such as an oral glucose tolerance test) and are interpreted as abnormally increased content of glucose in the blood (hyperglycemia). A specialist in the domain of medicine would generate the required knowledge to determine that this woman is suffering from a "gestational diabetes" and consequently inform her about what it involves and how she should manage it through diets, exercise or medication, if necessary. Finally, if the woman in question does not understand English, an equivalent term in another language (e.g. *Schwangerschaftsdiabetes* in German) is required to communicate this condition to her effectively.

2.1.2 Communication

The processes involved from observing data to generating knowledge through conveying meaningful information are embraced by the need for communicating and exchanging ideas.[2] When information needs to be transferred from one language to another one or more, interlingual[3] or multilingual communication is required.

The act of exchanging information between two or more persons is known as human communication. This can happen in a wide variety of ways and at many levels of sensory perception. From a communication theory perspective, communication happens when units of information are transferred from a source or sender to a target or receiver. In this process, information is represented within a context, coded in a message and transmitted through a channel by the sender. A successful transmission of information happens when the receiver is able to receive it, decode it and understand it. Although from a technical point of view, Shannon and Weaver's, (1949) statistical theory of signal transmission has usually been used as a landmark by many scholars to develop models of communication due to its simplicity. Shannon and Weaver's work focused on the physical problems involved in encoding and transmitting messages and the properties of the communication channels used for such transmission (i.e. radio, telephone, fax, etc.).

One of the main outputs of their theory was a model of the communication process (Fig. 2.2), which provided a basic framework for explaining more complex acts of communication, such as Nida and Taber's model (Fig. 2.3) including the translation function embedded between the sender and the receiver with the translator acting both as the receiver of the message in the source language and the sender of the message in the target language (Nida and Taber, 1969, p. 23). A more recent approach to this communication process in the Translation Studies field was that of O'Hagan and Ashworth, (2002), who drew on Weaver's model about a "Translation-mediated Communication"

---

[2] Actually, the word communication comes from the Latin word *communicare*, that means "to share," and more literally "to make common," from the Latin word *communis* (i.e. "shared by all or many") (Harper n.d.).
[3] Also called "translation proper" in seminal works (Jakobson, 1959, p. 233) or "translation-mediated communication" in more recent approaches to this concept in the digital era (O'Hagan and Ashworth, 2002, p. 1).

Information and translators 9

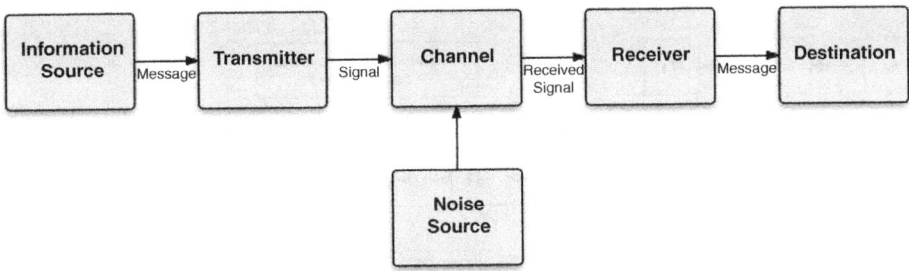

Figure 2.2 Shannon-Weaver's Model of Communication

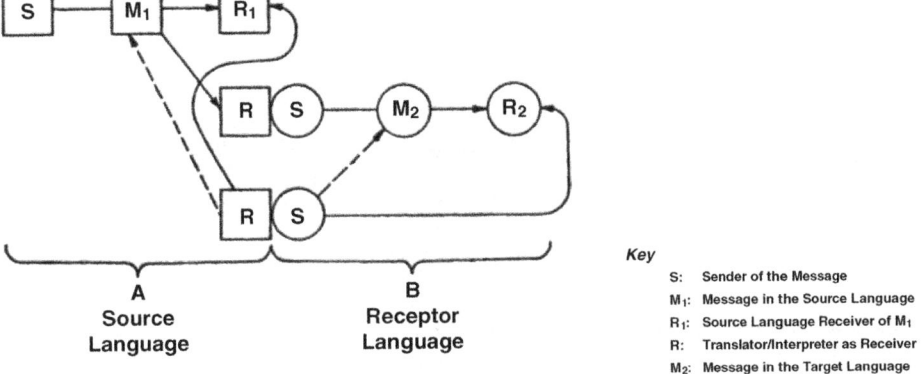

Figure 2.3 Nida and Taber's Model of Communication and Translation

(TMC) and combined it with the concept of "Computer-mediated Communication" (CMC) to explain their approach to today's digital communications environments, focusing on the interpreting setting of multilingual communication (Fig. 2.4).

Similarly, from the linguistic perspective, Jakobson (1971) described a model of the communication process including analogous elements, although focused on human communication. Jakobson's model (Fig. 2.5) distinguishes six communication functions (not displayed in this picture), each associated with a dimension of the communication process, namely 1. Referential, 2. Aesthetic/poetic, 3. Emotive, 4. Conative, 5. Phatic and 6. Metalingual.

Although these models have been criticised due to their linearity in the communication process and broader conceptions of communication theories have been developed to include social and behavioural aspects of the communicative act (see, for example, Bateson's discussion in Winkin, 1984), a simplified understanding of the concept of communication can be summarised from the models presented as conveying information between one or more sources and one or more receivers in a process that has the following characteristics (Beynon-Davies, 2002, p. 33):

- It involves two or more parties.
- One or more of the parties in a communications process will be the sender with intentions to convey.

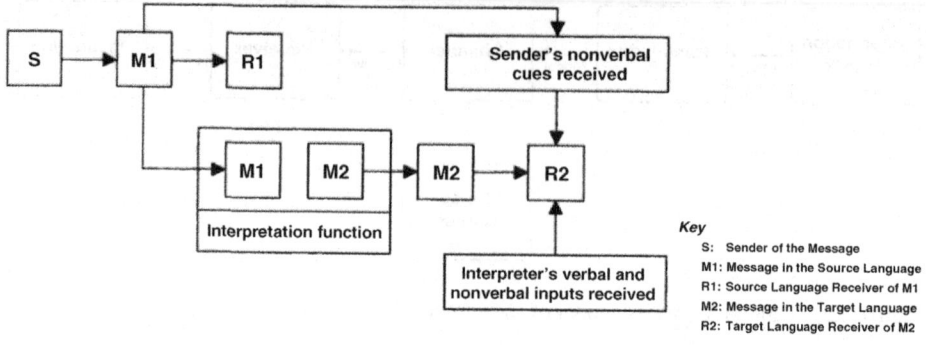

Figure 2.4 O'Hagan and Ashworth's Model of TMC

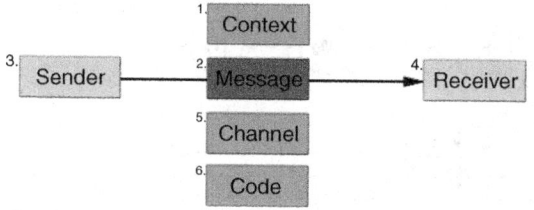

Figure 2.5 Jakobson's Theory of Communication

- The intentions of the sender will be expressed in a message.
- The message will be transmitted by the sender in terms of signals along some medium.
- One or more of the other parties will be a receiver. Receivers have the ability to interpret the signals.

If a particular day-to-day working context is taken, understood as providing referential conditions, as Jakobson's model shows, information is transmitted via both formal and informal communication (Bocij et al., 1999, p. 8). Certain pieces of information that are part of formal communication present data in a consistent and structured manner, such as database records, reports, guidelines and template-based documents, and therefore are likely to be accurate, relevant and serve a specific purpose. In a multilingual communication setting, for example, a translator might perform a terminological database query to obtain the equivalence of a particular term in another language. This might result in an output that presents one or more records with related information, such as definitions, equivalences in other languages, examples of use within a context etc. In this situation, if the translator finds the required information to understand the term and is capable of evaluating which equivalence will convey the same concept as the original message, this formal communication will suffice to the translator's purpose. In this example, the context is defined by the multilingual communication setting, between two different languages and cultures and within a domain of expertise; the sender is the translator, who places a query (the message) through a terminological database form (channel) using keywords, Boolean operators or a particular query language (code), and the receiver is the system that analyses the query to provide a result. However, if the output generated by the query does not provide the expected information or a deeper level of detail is necessary to make a decision about using one or another term in the target language, further communication by other channels might be required, like asking a domain expert by email or placing a post in an online specialised discussion forum.

Thus, the inflexibility of formal communication might not be suitable to deal with in-depth understanding required to an effective decision-making process or might not provide the information that enables the generation of new knowledge. On the other hand, informal communication, either transmitted by word of mouth or by any computer-mediated means, can be harnessed as a valuable resource to make a decision or generate knowledge. Informal communication can be much more flexible since it does not follow guidelines or patterns in the same way as formal communication, therefore the degree of detail can widely vary, although so can the accuracy depending on the source and the way of conveying the message. Given the diversity of channels and the complexity of the problems involved in this way of accessing information, informal communication can follow a rather laborious and manual process and therefore it might not be suitable to deal with large volumes of information or to provide fast and efficient communication.

2.1.3 Information systems

> We are surrounded by systems. Our bodies are made up of various systems, such as a digestive system and a central nervous system. We live on a planet that is part of the solar system. We engage with people in groups that form social, political and economic systems. We are educated in the use of number systems. Modern organisations would collapse without information systems.
> *(Beynon-Davies, 2002 p. 45)*

A system can be defined as "a collection of interrelated components that work together towards a collective goal [and which] function is to receive inputs and transform them into outputs" (Bocij et al., 1999, p. 25). In our modern information society, where information is continuously exchanged, and not only within people and organisations that speak the same language, but also between different cultures, countries and languages, there is an increasing need for multilingual communication. If this diversity in the delivery and dissemination of information is added to the very same diversity of the business activity, which involves different environments, different organisations and different sectors of economy, we end up with a rather complex understanding of how this multilingual communication can be achieved and how the businesses of this sector can successfully provide the services required from them. In fact, as Beynon-Davies, states (2002, p. 73) "even organisations in the same industrial sector will operate differently [...] to achieve something of a competitive advantage over their competitors in the market-place [...] in a number of ways: through differentiation in human activity, effectiveness in human activity and/or efficiency of human activity." Therefore, although there are core elements in the functions and ways to operate that most businesses have in common, each organisation, understood as a system, and more precisely as an information system, will necessarily be different.

This part of the chapter provides an overview of what is an information system to set the foundations of this type of organisational system and understand multilingual communication practice within an information systems approach.

There is an established and rich field of knowledge focused on information systems which "deals with systems for delivering information and communications services in an organization and the activities and management of the information systems function in planning, designing, developing, implementing, and operating the systems and providing services. These systems capture, store, process, and communicate data, information, and knowledge" (Davis, 2000, p. 62). The vast body of literature in this field covers the definition of basic concepts around information to the development of expert systems, namely,

- positioning of the field from informatics, organisational, and management perspectives;
- information and communication technology concepts (including hardware, software, telecommunications and networks);
- applications of information systems to e-commerce;
- types of information systems (including business information systems, management information systems, decision-support systems, executive information systems and intelligent and expert systems);
- systems' design, planning and development;
- information systems' strategy;
- security and ethical issues around information systems.

According to Davis, from a system-oriented perspective, an information system of an organisation consists of "the *information technology* infrastructure, *application* systems, and *personnel* that employ information technology to deliver information and communications services for transaction processing/operations and administration/ management of an organization. The system utilizes computer and communications hardware and software, manual procedures, and internal and external repositories of data. The systems apply a combination of *automation, human actions,* and *user-machine interaction*" (Davis, 2000, p. 67, emphasis added by the author).

Bocij et al. define an information system as "a group of *interrelated components* that work collectively to carry out *input, processing, output, storage and control actions* in order to convert data into information products that can be used to support forecasting, planning, control, coordination, decision making and operational activities in an organisation" (Bocij et al., 1999, p. 27, emphasis added by the author).

Beynon-Davies defines an information system in terms of a socio-technical system: "An information system [...] consists of both information technology and human activity" [...] "Information systems are forms of *human activity system* [...], systems of *communication* [that] involve people in producing, collecting, storing and disseminating information" (Beynon-Davies, 2002, p. 65, emphasis added by the author).

From the definitions above it seems clear that an information system is one that enables an interaction between information, technology, application software and users. Thus, a suitable working definition of this concept in broad terms could be the following: information systems are combinations of hardware, software and telecommunications networks that people build and use to retrieve, collect, classify, store, process and disseminate information to support administrators' decisions on how to control, develop and operate a business or an organisation.

What are the main components of such a system? If the elements that are mentioned in the definition (especially those emphasised in italic) are understood as parts that are put together to make it work towards a common end, they can be considered as the resources on which an IS relies, namely, people, hardware, software, communication and data.

- *People resources.* Include the users of an information system and those who develop, maintain and operate the system, from managers and end-users to technical support staff.
- *Hardware resources.* Include all the physical machines, including computers, peripherals, tablets, smartphones and other input and output devices, as well as the media used by them. They provide the processing, communications and storage capabilities required by application software systems and user activities. Traditionally referred to as the IT (information technology), along with software and, later on, with communication technologies (ICT). Due to the later developments in cloud computing or distributed computing,[4] it is important to note that physical does not equal to local, since resources and infrastructures can remotely be accessed and used through the net, including both machines (computers and virtual machines) and media (virtual disk storage and datacentres), thus providing enhanced collaboration, mobility, security and costs [see the concept of "Infrastructure as a service" (IaaS) (Voorsluys et al., 2010, p. 13)].
- *Software resources.* Include all the operating systems, computer programs, mobile applications and non-physical resources used by the hardware resources to control the operation of a computer system. Software resources are broadly classified as (Beynon-Davies, 2002, p. 128): system software (collection of programs which coordinate the activities of hardware and all programs running on the computer system, i.e. operating systems running on computers, tablets, mobiles, etc.); communication software (to enable communication between different computing devices in a network, i.e. standard protocols such as HTTP, FTP or TCP/IP); and application software (applications or packages of applications that run under a system software and that are designed to perform a particular set of tasks, i.e. general-purpose or specific-purpose programs and applications employed by users). Similarly to hardware resources, software can be locally installed on a computer or a device or can be accessed remotely [see the concepts of "Platform as a Service" (PaaS) and "Software as a Service" (SaaS) (Voorsluys et al., 2010, p. 14)].
- *Data resources.* Include all of the data to which an organisation or an individual have access. In a business setting, data are stored in repositories, formally organised as a database or a knowledge base, or informally stored in a number of files and documents. These repositories include all text and multimedia stores of documents, data search results, analyses, reports, e-mails, conversations, guidelines, procedures etc. that are required for performing the organisational activities.
- *Communication resources.* Include the networks, hardware and software required to support the transmission of data. In order to connect people to hardware, software, data and other

---

[4] The ability to run a program or application on many connected computers at the same time; or to use the hardware resources of computers and virtual machines that are accessed through a network and not physically located in the same place as the user (Mell and Grance, 2009; Voorsluys et al., 2010). A more comprehensive explanation of cloud computing can be found in Buyya et al., (2010).

people, computer networks must be enabled through channels for information that allow interrelating, sharing and collaborating among these elements.

Modern business practice heavily relies on computerised systems, formalised as information systems to a greater or lesser extent. From an IS perspective, organisations typically comprise different levels and functions of management that require different levels of information system capabilities and information to be effectively managed, namely, an operational level, a managerial/tactical level, and an executive/strategic level of management.

- The operational level of an organisation (also called low-level) uses information systems to improve efficiency by automating routine and repetitive tasks such as recording, summarising, sorting, updating or merging data, thus involving more structured information and decisions. Information systems at this level include Transaction Processing Systems (TPS) to record and process the data from business transactions, likely to be automatised or semi-automatised and processed in batches; Process Control Systems to automatically support and control manufacturing processes in engineering environments; and Office Automation Systems (OAS) to improve efficiency by applying ICT to common tasks, enhance communications and increase productivity.
- The managerial or tactical level of management (also called mid-level) of an organisation is focused on monitoring and controlling operational-level activities and information systems aimed at optimising the operational efficiency of the human, information and technical resources of an organisation at all levels of management by delivering reports of routine information. Decision making at this level involves semi-structured decisions because solutions and problems have to be made on aggregated or summarised data and are not clear-cut and often require judgment and expertise, and information systems under this level of management are referred to as Management Information Systems (MIS).
- The executive or strategic level of management (also called top-level) uses information systems designed to support the highest level of management. These systems are called Executive Information Systems (EIS) and are used to retrieve and manage summarised information to avoid an information overload, so that it can be interpreted and used to support the achievement of an organisation's business objectives or its long-term strategic goals.

Even though organisations have usually followed this traditional pyramidal and hierarchical style of management levels, it is also true that non-hierarchical management structures and new organisational forms are becoming more usual today (Pettigrew et al., 2003; Kastelle, 2013). In addition, self-managed small organisations or freelance professionals usually have a much simpler structure or gather roles and functions around one or a few people. These realities, together with more global approaches to information systems, increased levels of collaboration within large organisations, and information systems operating with a wider functional scope, have originated a wider conception of information systems that span all levels of the organisation (see Fig. 2.6).

If operational-level IS are aimed at "getting things done" to meet day-to-day business demand and MIS are focused on "doing things right" to give companies a competitive advantage, a broader-scope type of information system that ensures that all this happens and the right decisions are being made are Decision Support Systems (DSS). These information systems are focused on providing support to problem-specific

Information and translators   15

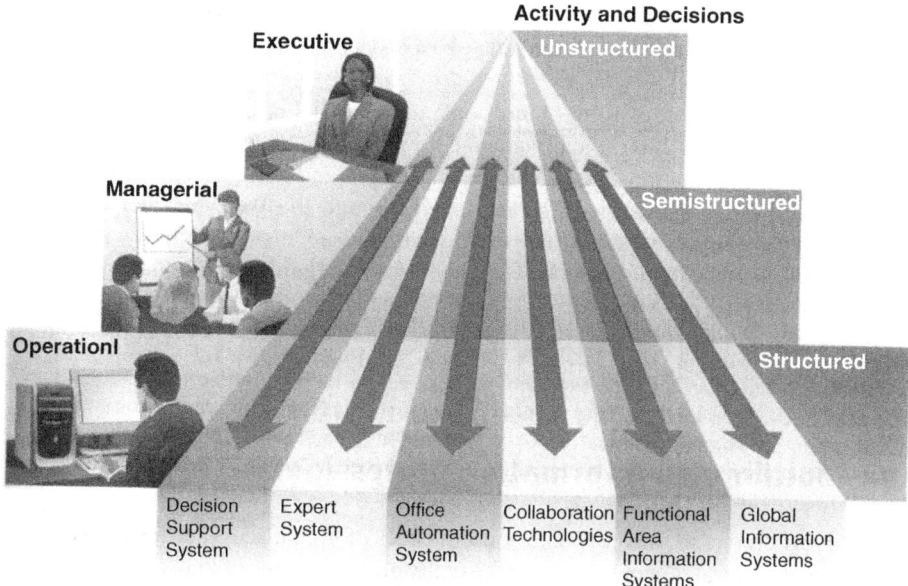

Figure 2.6 Organisational boundary-spanning information systems
(Jessup and Valacich, 2005, p. 219)

decision making and problem solving to optimise the effectiveness of an organisation, i.e. "do the right thing" (Stair and Reynolds, 2011, p. 436). Although, as it happens with MIS, they are primarily designed to assist the top management function of a company, decisions are made at all levels of an organisation operation, thus becoming a useful tool for those organisations that need to deal with information and knowledge in an intensive, interactive, and flexible way.

Stair and Reynolds (idem, p. 438) define the following functions of a DSS:

- handle large amounts of data from different sources (e.g. databases, data warehouses, networked systems);
- provide report and presentation flexibility (e.g. different formats and outputs);
- offer both textual and graphical orientation (e.g. text, tables, charts, drawings);
- support drill down analysis (e.g. information at several levels of detail: project > phase > activity > task);
- perform complex, sophisticated analysis and comparisons using advanced software packages (e.g. bring stand-alone analyses together in the same DSS);
- support optimisation, satisficing, and heuristic approaches (e.g. "what-if analysis" to enable hypothetical changes to problem data and observing the impact on the results, "goal-seeking analysis" to determine the problem data for a given result, or "simulations" to duplicate the features of a real system.

When functions can be automated to some extent and solutions can be directly presented to decision makers, IS are called expert systems. In addition, given the complexity of DSS a number of advanced types of IS can be considered as supporting decisions, such as these expert systems, artificial intelligence systems, neural networks, fuzzy logic, data mining, knowledge-based systems and intelligent knowledge-based systems.

Going back to the central point of this book, multilingual information management, technology is being used to radically change how the translation business is conducted, from the way translations are produced, distributed and accounted for, to the ways translation and language services are marketed through the Web. A wider than ever range of ICT is available to support this knowledge-based business activity. However, there is a lack of attention to these businesses from the organisational, managerial and information systems perspectives, and the impacts on information systems in translation businesses settings have not been much investigated. The rest of the book is aimed at contributing in this direction.

2.2 Multilingual information professionals

As stated earlier, one of the inevitable consequences of globalisation is the great deal of linguistic and cultural diversity found in virtually all the spheres of knowledge in which information needs to be communicated. Companies, institutions and anyone who needs to offer services and products across the globe has to cope with multilingual information and require professional services that ensure quality and reliable work. Thus, this kind of solution should be provided by qualified professionals in possession of specific competences and knowledge involving linguistic and cultural communication, translation skills, acquaintance with text genres, cultural awareness, domain-specific expertise, information research and management, and interpersonal and instrumental capabilities, to name but the main ones (Kelly, 2002; Hurtado Albir, 2007).[5]

The concept of competence is a rather complex one that arises from the competence-based training pedagogic approach and entails much more than just possessing knowledge or a skill. It has deserved considerable attention by scholars, particularly since the emergence of the European Higher Education Area (EHEA). One definition in this context, provided by the European Commission Working Group "Basic Skills, Entrepreneurship And Foreign Languages," is "to refer to a combination of skills, knowledge, aptitudes and attitudes, and to include disposition to learn as well as know-how" (European Commission, Directorate-General for Education and Culture 2003, p. 10). Another widely cited and used definition due to its comprehensiveness is the one proposed by Lasnier (2000, p. 32): "Une compétence est un savoir-agir complexe résultant de l'intégration, de la mobilisation et de l'agencement d'un ensemble de capacités et d'habiletés (pouvant être d'ordre cognitif, affectif, psychomoteur ou social) et de connaissances (connaissances déclaratives) utilisées efficacement, dans de situations

---

[5] These two recognised scholars in the Translation Studies area provide a comprehensive review of the translation competence and the models that have tried to explain the sub-competences behind this professional profile, i.e. what specific knowledge, skills to solve problems and social skill translators must hold. As this discussion falls beyond the scope of this book, a detailed reading of these two works is highly recommended for a thorough account of this matter.

ayant un caractère commun" and translated into English by Hurtado (2007, p. 166) as: "A competence is a complex *know how to act* resulting from integration, mobilization and organization of a combination of capabilities and skills (which can be cognitive, affective, psycho-motor or social) and knowledge (declarative knowledge) used efficiently in situations with common characteristics." In addition, the competence-based training approach distinguishes between those competences that are closely related to a discipline (specific competences) and those that can be applied to all disciplines (general or transversal competences).

Professionals working with multilingual information do not only include translators per se, but also a range of professional profiles that provide language-related services and act as information facilitators among different languages and cultures, namely, interpreters, cultural mediators, language editors and language services project managers. All these profiles share the specific competences stated above, to a greater or a lesser extent, depending on the particular context in which they develop their professional activity. For this reason, the term "Multilingual Information Professional" (MIP) can be a suitable hyperonym that embraces this wide range of professionals, and even other related ones such as terminologists, technical writers or documentalists. On the other hand, translators being the main community of practice that deals with multilingual information among all these MIPs and having a rather consolidated academic discipline, it does not seem odd that this particular profile has been more thoroughly researched by scholars (c.f. Chapters 3 and 4). On many occasions, studies and existing research are focused on translators in particular, but in other cases discussions deal with issues, activities or contexts that are not limited to translators and could be applied to other MIPs. Thus, for simplification purposes and to refer to most of the professionals researched by previous studies, the term "translators" is used throughout this book to refer to both specific professionals that mainly work as such and multilingual information professionals in general, unless specified.

2.2.1 The role of the translator

If the figure of the translator is then taken as a paradigm of a MIP, a brief introduction into what being one involves and how they interact with the process of interlingual and intercultural communication can help to understand the needs of this type of professional. Schäffner proposes a definition of the role of the translator, based on an article from Schmitt, (1998), which offers a clear, comprehensive and current view of what is a translator, highlighting in particular the skills required of a modern translator, and his or her orientation towards meeting translation market challenges.

> *Translators (and interpreters) are experts for interlingual and intercultural communication, and assume full responsibility for their work. They have acquired the necessary professional expertise, above all linguistic, cultural and subject-area competence, and are equipped with suitable technological skills to meet the challenges of the market today and those to be expected over the coming years. On the basis of source material presented in written, spoken or multi-medial form, and using suitable translation strategies and the necessary work tools, they are able*

to produce a written, spoken or multimedial text which fulfils its clearly defined purpose in another language or culture. Translators are engaged in fields ranging from scientific and literary translation over technical writing and pre- and post-editing to translation for stage and screen.

(Schäffner, 2000, p. 25)

Since ancient times, even before St. Jerome translated the Bible into Latin, the essence of what a translator does has not changed very much. In Translation Studies, the process of the translators' work has traditionally been divided into three main stages, namely, pre-translation, translation and post-translation (Holmes, 1988; Hatim and Mason, 1990; Austermühl, 2001). Translators are usually seen as persons who have a good command of at least two languages, maybe a degree in linguistics, a languages-related degree, or studies in translating and interpreting. It is obvious that linguistic skills are very important for translators, but there are other skills that must also be acquired by a translator, such as expert knowledge of a specialist subject, cultural and communication competence, or technological skills (Schmitt, 1998). This is particularly important in specialised contexts, where translators are usually seen before the eyes of experts in a specific field of knowledge as outsiders to their professional community and must be acquainted not only with text genres that help them to understand the mechanisms behind this particular act of communication, but also with the interlingual and intercultural conventions required to enable communication in specialised multilingual contexts (García Izquierdo and Borja Albi, 2009, p. 9). Essentially, the core tasks of the translator's role have not changed much over time:

- Translators have always had to draft the translated text, and proofread it, before they could deliver it (i.e. take care about the production of documents and translations and quality control tasks).
- Translators have always had to look for answers to terminological or subject-knowledge problems (i.e. take care about information search tasks).
- Translators have always had to deal with clients to receive the source material, send it back, and invoice them (i.e. take care about communication and business administration tasks).
- Translators have always had to promote themselves as language professionals capable of undertaking translations and show their potential clients that they are suitable professionals to perform their translations (i.e. take care about marketing themselves).

As indicated earlier, the number of translators operating on a freelance basis has substantially increased over the past decade (Fraser and Gold, 2001; Holland et al., 2004, p. 254; Locke, 2005, p. 19), and being a freelance translator also means taking on additional tasks. Freelance translators do not only have to undertake translation assignments, but also have to deal with their clients, manage their translation assignments, sort out payments, acquire new ICT equipment, continue professional training, and other activities that are necessary to run a translation business. In other words, in their day-to-day life as translators, they have to assume a number of activities that are part of running their translation business, but that are not part of the core translation task. These requirements of the freelance translators have been reported by authors such as Varona, (2002, p. 202) who cites budgeting, pricing, customer services or marketing functions, and Locke, (2005, p. 50) who adds hardware and software acquisition to the examples of activities the freelancer must undertake. In this sense, freelance translators are not only language professionals capable of providing translations, but also professionals who have to run their own business and take decisions upon which business and ICT strategies they have to follow to remain competitive in an ever demanding market.

In addition to the added tasks originating from freelancing, the tools and resources used to support the translator in the tasks performed have also changed over time.

Technology and translation

As stated in the previous chapter, today's multilingual and globalised society is in high demand for translation services that facilitate communication between different languages and cultures, and deliver documents and services meeting the quality standards of target markets (Taravella and Villeneuve, 2013). This increased need for multilingual communication has been exacerbated by a number of factors, including an emphasis on globalisation and international trade by the business community (Lange and Bennett, 2000, p. 203), the advent of the World Wide Web as an international marketing tool, the ever-growing amount of digital content such as web-based ones, or the rise of the software localisation industry (Sprung, 2000, p. ix). In multilingual regions like Europe, the forging of closer trading relationships between countries and the enlargement of their political bodies (i.e. the European Union) have highlighted an awareness of the need for multilingual communication, and again fuelled demand for translation services (Roxburgh, 2004).

Along with the growth faced by the translation market (Sprung, 2000), more and more clients are requesting faster, better and cheaper translation services. Schäffner, (2000, p. 7) indicated that "translations need to be done ever more quickly, much more efficiently, and at a high quality." Client demand has therefore meant that the language services sector, especially the translation sector, has had to develop innovative production processes and software tools to lower transaction costs, work faster and provide consistently high quality (Shadbolt, 2003).

In addition, the increasing availability of personal computers and the prevalence of electronic tools over handicraft ones has also facilitated the development of information and communication technologies specifically designed for professional translators. The development of technologies such as computer-aided translation (CAT) tools,[1] the main component of which is based on translation memory (TM) technology, has allegedly led to significant increases in the quantity (productivity) and quality (efficiency and effectiveness) of translators' work (Heyn, 1998; Somers, 2003c), and these tools have been deemed to be one of the most useful facilities for translators (Hutchins, 2005a).

Running in parallel with the increasing demand for translation services and the availability of specialised ICT for translators, various organisational developments have had, and are indeed continuing to have, a considerable impact on the translation services sector. For example, many in-house translation departments have closed as large commercial organisations have found it necessary to downsize and focus on core competencies in order to reduce costs (Fraser and Gold, 2000, p. 3; Locke, 2005, p. 19). As a result of this divestment, organisations now tend to outsource more translation

---

[1] This concept is defined in the following section of this chapter.

assignments to freelance translators. Public sector organisations have adopted a similar approach and now tend to rely on the services of freelancers, in conjunction with a core body of in-house translators. As a result of these developments, a substantial proportion of translators, in the UK and elsewhere, now work on a freelance basis (Holland et al., 2004, p. 254; Locke, 2005, p. 19).

Technological developments in the freelance translation sector have provoked much discussion among translators at professional conferences and seminars, as well as via online discussion groups, but the adoption of specialist tools, such as the above mentioned CAT tools, did not receive much attention from the academic community until the end of the 1990s and had not started to be thoroughly investigated until the first decade of the 21st century. Since the beginning of this century, it has been claimed that translation professionals have had to catch up with the increased demand for translation services and that, to do so, translation memory and terminology management solutions – the main features of CAT tools – should not only be used by large multilingual services suppliers, but also by small translation companies and freelance translators (see, for example, Joscelyne, 2003). Joscelyne's statements were based on the findings of the surveys conducted by the Localization Industry Standards Association (LISA)[2] (Lommel, 2002; 2004), which reported a growth in the use of translation memory technology and evidence of translation companies becoming more productive due to such use. Research has generally concentrated either on evaluations of CAT tools' technical features (see, for example, Weßel, 1995; EAGLES, 1996; Whyman and Somers, 1999; Esselink, 2000; Austermühl, 2001; Zerfass, 2002; Bowker, 2002; Quah, 2006), or has tended to be focused on the working environments of in-house translators (King, 1998; Blatt, 1998; Chanod, 1998; Rinsche, 2000; Lange and Bennett, 2000). While some of the latter have been more comprehensive in their coverage of translators' working practices and the technology used, their findings are inevitably now somewhat dated as the studies were undertaken prior to, or in the very early days of the widespread commercial availability of CAT tools. Such studies also include Smith and Tyldesley (1986), Fulford et al. (1990), Fulford, Höge and Ahmad (1990), and a European study, carried out as part of the LETRAC Project, undertaken by the end of the nineties, and reported in Reuther, (1999).

In addition, discussions about CAT tools have, at times, been emotionally charged, primarily because of the threat to job security which some translators fear computer-assisted aids pose to the translation profession (see for example, Shields, 1999; Fenner, 2000). A number of concerns about the use of CAT tools such as low job satisfaction levels due to the use of CAT tools, the unsuitability of CAT tools for freelancers' needs, the high cost of the tools, or conservative attitudes towards technological investments have been mentioned in existing literature (see for example, Fulford et al., 1990; Heyn, 1998; Hutchins, 1999; Esselink, 2003).

---

[2] The Localization Industry Standards Association (LISA) was a non-profit organisation set out to develop standards and specifications for information exchange in software localisation and that remained active between 1990 and 2011. It played a significant role in developing localisation and terminology standards, such as ISO 30042:2008 (Systems to manage terminology, knowledge and content – TermBaseeXchange (TBX) (ISO/TC 37/SC 3 2008)).

Evidence regarding the uptake of CAT tools by freelance translators, the benefits of using these tools, and the problems associated with their use, has only started to be reported by research over the past ten years or so and is discussed in more detail in section 3.3 of this chapter.

In the following sections, a review of the range of ICT available to translators is introduced by an overview of the development of translation tools, their evolution over time, the issues around the use of CAT tools by freelance translators, and a discussion of the "translator's workstation" concept through the models of translation tools suggested by previous research.

3.1 Tools to support translators

Traditionally, a translator just used paper and ink to write, and paper dictionaries and libraries to do research. As technology evolved, the use of dictation machines and typewriters (mechanical and later electronic) assisted translators in their work. However, it was the proliferation of microcomputers – personal computers (PCs) – that formed a turning point in the way that translators work. The mere use of word processors greatly assisted translators in tasks such as revising and editing translations or adding format to documents. The development of computer-based reference works on electronic media, such as CD-ROM/DVD first, and then the advent of the Internet and electronic communications, multiplied the resources that translators could use in order to increase their productivity and quality of their work, and to improve the ways they communicate and transfer information.

Many different terms are used when referring to computer-based tools and resources that support the translation process, for example, "translation software" (Hutchins, 2000a), "translation tools" (Esselink, 2000; Langewis, 2002), "language technologies" (Shadbolt, 2003), "electronic translation tools" (Austermühl, 2001), "machine-aided translation" (Quah, 2006), and "translation environment tools" (TEnTs) (Zetzsche, 2006). The term "computer-aided" or "computer-assisted translation" tools is also used to refer to the computer-based applications that support the translation process (Hutchins and Somers, 1992) and seems to be the one that is more familiar to scholars of the Translation Studies field, professionals and industry (Quah, 2006, p. 6).

In view of the existing terminological variations, for the purpose of the present work, the term "translation tools" is used to refer to the different terms given to all the computer and Internet-based resources supporting the translation process. The term "CAT tools" is used to refer to the specific set of computer applications designed to assist translators in producing faster and consistent translations, by storing source and target language pairs of text segments, such as sentences or paragraphs of previous translations, and by retrieving exact or partly similar equivalences of text segments with the source and already translated text, during the production of new translations. The equivalences suggested by CAT tools occur at a term level (through terminology management functions) or at a longer segment of text level, such as a sentence or a paragraph (through translation memory functions). Those functions, along with other ones usually included in CAT tools (e.g. document alignment, word count, file format filtering, project management) have been designed to help translators during the core activities in their work, i.e. translation production and storing and retrieving terminological information. Next, a brief definition of the two main components of CAT tools is provided:

A "translation memory" (TM) can be defined as a "database that stores previously translated sentences that can be retrieved in future translation projects in an attempt to prevent [eliminate] repetitive, time-consuming work. Pre-translated sentences in the text are retrieved via fuzzy [approximate] matching, leaving only parts of the sentence that do not have matches to the translator" (Tunick, 2003, p. 14). This type of application is the main function of the CAT tool packages available today and is one of the most useful facilities for translators (Hutchins, 2005b, p. 13).

A "terminology management tool" or "terminology management system" can be defined as "a program that catalogues words and phrases along with pertinent related information [e.g. grammatical, context] in a database in a manner conducive for use in linguistic applications" (Langewis, 2002, p. 6). The main purpose of this use in the case of translators is to store and retrieve terms in order to ensure a greater consistency and to improve quality in the use of terminology during a translation project, across teams of translators, or through a same client or field of knowledge. Yet, the linguistic application of terminology management tools also involves a wider range of functions that also contribute to the management of multilingual information, such as the automatic recognition of terms when integrated with a word processor or a translation memory, "essentially a type of automatic dictionary look-up" in words of Bowker, (2002, p. 81); term extraction functions, thus facilitating batch pre-translation tasks and post-editing or quality assurance terminology check-ups (idem, p. 82); or other uses for managing terminology in machine translation, termbases, knowledge-based systems or corpus-related applications. A detailed account of terminology management applications can be found at the most comprehensive work about this topic, *Handbook of Terminology Management*, particularly in volume II (Wright and Budin, 2001). In a more recent contribution, the expert linguist Alan Melby presents an updated view of terminology management in the context of the latest developments, including an optimistic prediction about the relevance of termbases in future scenarios (Melby, 2012).

3.2 Translation tools: origins and evolution

The historical development of computers for the translation of human languages has been well researched and documented by scholars (see, for example, Melby, 1982; Slocum, 1988; Hutchins and Somers, 1992; Melby, 1992, 1998; Hutchins, 1996; Kay, 1997; Abaitua, 1999; Hutchins, 2001b, 2001a, 2002, 2014). John Hutchins' work deserves, at the least, a separate mention since very few scholars have probably performed such a lifetime careful effort to archive and document almost every advance in this field published in English since 1980,[3] which still remains active today and can be consulted at his "Machine Translation Archive" at *http://www.mt-archive.info*.

---

[3] After the update of 18 January 2014, the archive contained over 10,550 references of articles, books and papers on topics in machine translation, computer translation systems and computer-based translation tools.

The origin of research on using computer aids for translating natural languages can be attributed to Warren Weaver of the Rockefeller Foundation, who was one of the pioneer researchers who put forward the use of cryptographic techniques, the application of the Claude Shannon information theory and statistics, and speculations about universal principles underlying natural languages.

Early developments aimed to achieve automatic ways of translating texts from one language to another, using what was called "Machine Translation" (MT). An MT system can be defined as "software for automatic translation, where input units are full sentences of one natural language and the output units are corresponding full sentences of another language" (Hutchins, 2000a, p. 4) without the intervention of any human translator (excluding pre-editing or post-editing) (Slocum, 1988).

The first public demonstration of an MT system was held in 1954, in a collaboration of IBM with Georgetown University. These early systems consisted basically of large bilingual dictionaries and a set of rules that allowed the system to determine the syntactic order of the output. This initial optimism made it possible to think of developing a system that offered fully automatic high quality translation (FAHQT). In this context of euphoria, two systems that are still used in the present, *Systran* and *Metal* were developed by Georgetown and Texas Universities.

After one decade of optimism and investments supporting predictions of successful MT systems, the outputs produced were still disappointing, and the human translator always had to be present to widely revise (post-edit) the outputs, so they failed to fulfil the expectations created. By 1964, the US government sponsors had become increasingly concerned at the lack of progress, and the US National Research Council set up the Automatic Language Processing Advisory Committee (ALPAC), which concluded in a 1966 famous report that MT was slower, less accurate and twice as expensive as human translation, and that there was no immediate or predictable prospect of useful machine translation (ALPAC, 1966). It saw no need for further investment in MT research, and instead it recommended the development of machine aids for translators, such as automatic dictionaries, and continued support of basic research in computational linguistics.

Although the ALPAC report had great impact elsewhere in the Soviet Union and Europe, the drastic effect on MT research in the United States did not reproduce in Canada, France or Germany. In Europe, the Commission of the European Communities adopted the system *Systran* and sponsored an ambitious project called *Eurotra* in 1976. In the same year, another successful MT system for translating weather reports, *Meteo,* was developed in Canada by the research group TAUM (Traduction Automatique de l'Université de Montréal). Despite new efforts in research during the 1980s, the expectations in machine translation success were less and less supported by governments, companies and institutions.

In the absence of successful results, there was a shift from fully automated machine translation (FAMT) systems towards the research and development of computer tools that assist translators, called Computer-Assisted Translation (CAT) tools (Hutchins and Somers, 1992), as in some way had been suggested by the ALPAC report (ALPAC, 1966) and Kay, (1997), in an internal working paper written for Rank Xerox in 1980, but not published until 1997. For 30 years, software developers had been trying to replace human translators with machine translation, but finally technological developments had started to focus on providing specific-purpose tools to these professionals: "This legacy has caused our industry [translation tools] to fall behind in performance improvements made in other [software] industries" (Hunt, 2003). While FAMT systems were based on rules, Rule-based Machine Translation (RBMT), the main new MT developments in the early 1990s were based on statistical methods or analogies, Analogy-based Machine Translation (ABMT), also called Example-Based Machine Translation (EBMT), which used compared translation corpora and no syntactic or semantic rules in the analysis of texts or in the selection of lexical equivalents.

Still lagging far behind CAT tools, which experienced an important increase thanks to the localisation industry, MT systems were deemed a fast real-time solution for a rising crowd of Internet users, in spite of their non-publishable quality results, leading to the introduction of MT online services with the launch of Babel Fish (Yang and Lange, 1998; Gaspari, 2004; Gaspari and Hutchins, 2007). Since then, FAMT has mostly focused on Statistical Machine Translation (SMT) methods to improve its results and spread its areas of application, mostly through the use of controlled language to pre-process MT input and post-editing MT output, but also to enhance CAT tools (Melby, 2012, p. 11).

For some time now, the development of CAT tools has been focused on combining TM and MT systems to try to get the best of both worlds and boost TM automatic segment retrieval by using MT methods (Melby, 2007; Kanavos and Kartsaklis, 2010). This has resulted in two scenarios: on one side, a number of CAT tools that have started to include MT capabilities (Heyn, 1996; Lagoudaki, 2008, p. 263), and on the other, post-editing MT output is becoming a more and more common practice in certain circles of professional translation (de la Fuente, 2012; Yuste Rodrigo, 2013).

MT efforts aside, whether too idealistic and far from meeting their ambitious expectations in the 20th century or more realistic through improved methods and gaining ground as aids to human aided translation, CAT tools represented the main advance in the ICT available to support the translator's work, and their availability to freelance translators increased the prospects of a widespread use of these tools (Heyn, 1998; Joscelyne, 2003; Somers, 2003a).

Practitioners have claimed that significant benefits can be obtained from their use of CAT tool packages, especially with regard to the reductions in the time invested in the translation process, and in the cost of their translation projects provided by the translation memory functions. Some examples of these claims can be observed in the following quotes:

> "The main benefit of tools is leveraging translated text from TM and reducing project management time and engineering support. [...] If we can reuse 40% of the weekly average output in leveraging TM, it saves us (and our clients) a lot of money. The larger savings, however, is in the form of project time scales. Tools and workflow can save at least as much in reductions of real-time use of project managers and engineers. The client benefits from the reduction in time-to-market – which is probably more relevant" (quoted by Hedley Rees-Evans in Shadbolt, 2003, p. 6).

> "TM software can reduce the length of the translation process by 50%. Additionally, reductions in total translation costs of between 15% and 30% can be realized" (Tunick, 2003, p. 14).

In these statements, a number of benefits of using CAT tools are highlighted, especially with regard to reutilising a significant portion of the translation output, and reducing the length of the translation process, which results in a reduction of costs for translators and their clients. However, they were made in the context of large translation services providers, and it remained unclear whether the same benefits are being obtained by freelance translation businesses. Studies such as those focused on freelance translators, as stated at the end of the introductory text of this chapter, have started to shed light on these issues, even though some scholars have already stated that TM-based CAT tools are "reaching [their] use-by date" (Garcia, 2009, p. 199) since they seem to be incapable of coping with the increasing translating needs of today's digital age, in favour of an MT post-editing role of translators (idem, p. 210) and forecasting a likely worsening of translators' working conditions.

Although Garcia's prediction does not leave human translators out of the game (yet), as he states that "[f]or now and the foreseeable future, stand-alone, unassisted MT is not yet the solution" (idem, p. 206), it was not the first time to state that translators' work will soon be done by machines, nor will be the last time to predict the end of human translators (as the trained multilingual information professionals they are today).

As a paradigmatic anecdote, in a recent interview with Alan Melby conducted by Jost Zetzsche (2013), he told that in 1984 he was approached by the director of the Eurotra machine translation project (one of the main European projects on MT mentioned earlier) at a computational linguistics conference, who told him: "Professor Melby, you are wasting your time developing translator workstations. Within five years there won't be any more human translators." Melby stood his ground in favour of human translators aided by computers and the idea of a translator's workstation, which not only became real, but also was the core idea underpinning CAT tools development, still being the most significant translation tool today. This concept of an integrated electronic environment, represented by the broad conception of "CAT tools" and of the "translator's workstation," is precisely the central point of the next section of the chapter.

3.3 The translator's workstation

There is a wide range of computer-based applications that has been developed to suit the needs of translators, i.e. translation tools, according to the explanation provided in section 3.1. Previous research has categorised translation tools in different ways. Here, some of the models for categorisation of translation tools are presented, along with some of the basic terminology used in the discussion about translation tools in general.

Before the appearance of CAT tools, Melby (1982) presented a three-level classification of computer-based tools, based on a functional approach (Fig. 3.1) that made up a "translator's workstation," a term widely used by experts in the field (Melby, 1992; Hutchins, 1998) to refer to the computer software and hardware used by translators.

One of the most popular classifications arises from the model presented by Hutchins and Somers (1992), which shows the degree of human involvement in the translation

Level 1	Text Processing Telecommunications Software Terminology Management Systems Others (DTP, Converter)
Level 2	Text Analysis Automatic Dictionary Look-up Bilingual Text Retrieval Other (SGML)
Level 3	Machine Translation

Figure 3.1 Melby's translator's workstation (based on Melby, 1982)

process. Figure 3.2 shows the dimensions identified by these authors. They differentiate between purely human translation and fully automatic high quality translation (FAHQT). The categories in between refer to tools that require some interaction between computers and human translators; either by using tools that aid professional translators (e.g. grammar and spelling checkers, online and CD-ROM electronic dictionaries), labelled as Machine-Aided Human Translation (MAHT), or by using semi-automatic translation tools that require human intervention for pre-editing and post-editing tasks, labelled as Human-Aided Machine Translation (HAMT). The authors use the term Computer-Assisted Translation (CAT) to refer to these two types of interaction.

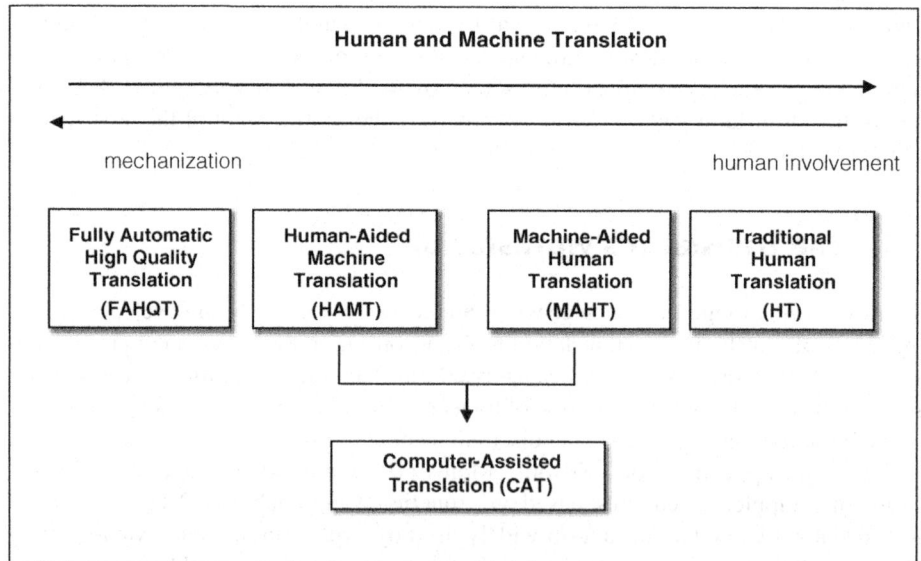

Figure 3.2 Hutchins and Somers' dimensions of translation automation (Hutchins and Somers, 1992, p. 148)

Technology and translation

	INFRASTRUCTURE	
	TERM LEVEL	**SEGMENT LEVEL**
BEFORE TRANSLATION	• Term candidate extraction • Terminology research	• New text segmentation, previous source-target text alignment, and indexing
DURING TRANSLATION	• Automatic terminology lookup	• Translation memory lookup • Machine translation
AFTER TRANSLATION	• Terminology consistency check and non-allowed terminology check	• Missing segment detection and format and grammar checks
	TRANSLATION WORKFLOW AND BILLING MANAGEMENT	

Figure 3.3 Melby's eight types of translation technology
(Melby, 1998)

In a later contribution of Melby, (1998), he categorised types of translation tools according to the traditional three phases of translation (i.e. pre-translation, translation and post-translation), and depending on the linguistic level in which the translator receives support, i.e. at individual term level, or at a whole segment level. Figure 3.3 shows the eight categories identified in Melby's model.

In an attempt to define categories of translation tools, the International Association for Machine Translation (IAMT) categorised them in two main groups (Hutchins, 2000b): automatic translation (MT) systems and computer-based translation aids. In this second wide group, those tools that are familiar to the general public (described as systems that provide linguistic aids for translation by Hutchins) are differentiated from those that have been specifically designed to support translators (translation support tools). Figure 3.4 below lists all the categories and the tools belonging to each one, according to the IAMT Certification group classification.

Automatic translation systems	Computer-based translation aids	
	Linguistic aids for translation	**Translation support tools**
Basic level systems ("entry level" or "home use") Standard level systems ("professional level") Advanced level systems ("company level")	Dictionaries (bi/multilingual) Language aids (grammatical) Spelling checkers Style checkers Terminology aids Specialised glossaries (areas/clients)	Electronic dictionaries Terminology management systems Translation memory systems Foreign language authoring systems Translator workstations

Figure 3.4 IAMT Certification initiative classification
(based on Hutchins, 2000b)

Automatic translation systems	Online translation systems	Translation support tools	
Home use MT system Internet/Web MT system Professional use MT system Client/server MT system	MT services (i.e. translation service via Internet or cellular) MT portals (i.e. access to MT services on Internet and/or to information about MT systems)	Electronic dictionaries Terminology management systems Translation memory systems Translator workstations	
		Alignment tools Pre-editing tools Localisation support tools	Added tools

Figure 3.5 *Compendium of translation software* classification of translation tools (based on Hutchins, 2000a)

There was a parallel, but independent, effort to compile a general guide to commercially available systems by Hutchins (2000a). In his compendium of translation software, he suggested a slightly different set of definitions which appeared to be more easily applied in practice. Additionally, Hutchins' compendium includes a greater number of translation support tools, does not include the category "Linguistic aids for translation" inside the broad category of computer-based translation aids, and adds another sub-category of MT systems, namely online systems. Figure 3.5 presents a summary of the categories of translation tools used by Hutchins in his work.

Austermühl (2001) reviewed some of the existing models (Melby, 1982; Hutchins and Somers, 1992) and proposed a more holistic process-oriented approach to understand that translation tools are an integral part of the translation process, and instruments that support the translator during the various sub-processes of translation. He states that in order to make an effective use of translation tools, the translator needs to determine what types of translation tools are needed at what stages in the translation process. In addition, he outlines that to use translation tools effectively and design and evaluate them, compatibility between the tools and the steps must be ensured. In the whole translation process, Austermühl differentiates three levels at which translation tools may help translators, namely translation workflow management level, linguistic and cultural transfer level, and automation level. Figure 3.6 below summarises the tools and resources that, according to Austermühl, provide support to translators during the sub processes at each level of the translation process.

All the classifications presented above are a valuable contribution to the translation studies research, offering classifications of different natures: the human-machine involvement in the translation process (Hutchins and Somers, 1992), the translation tools used at each phase of the translation process (Melby, 1998), the types of application at each sub-classification of translation tools (Hutchins 2000b; 2000a), or the types of application at each process and sub-process of the translation process (Austermühl, 2001). However, all these classifications focus either on linguistic

Technology and translation

Level	Sub-processes		Translation tools/resources
Translation workflow management	Client-translator communication		Internet-based communication tools, e.g.: - email - FTP - WWW

Linguistic and cultural transfer	Reception phase	Retrieval of background knowledge	- Encyclopedias
			- Knowledge databases
			- Information retrieval systems
			- Contacts to domain experts
			    - Mailing lists
			    - Newsgroups

		Source text analysis	- Terminology extraction tools - Terminology databases
	Transfer phase	Retrieval of linguistic, encyclopaedic and intercultural knowledge	- Electronic dictionaries (CD-ROM or Internet) - Terminology databases - Hypermedia systems
	Formulation phase	Syntagmatic relations and collocations	- Electronic dictionaries (CD-ROM/Internet) - Terminology databases - Style guides - Collocational dictionaries - Text corpora (CD-ROM/Internet)
		Document management process	- Alignment tools - Translation Memory tools - Terminology management tools
Automation	Whole translation process		Translation memories Software localisation tools Machine translation systems

Figure 3.6 Austermühl's process oriented approach
(based on Austermühl, 2001)

processes involved in the translation process, or on types of application that fit into a number of sub-categories of translation tools. These categorisations have a number of limitations from the perspective of freelance translators, since they do not include ICT that supports them within the context of a freelance translation business. Types of application which might support freelance translators' activities in other processes around the core translation function supported by CAT tools, such as the ones mentioned earlier in the chapter (i.e. financial management of their translation business and promoting their services) would not fit into these models of ICT adoption. Therefore, from the translation industry point of view, existing models do not fully consider ICT in the context of freelance translators, and thus these classifications are restricted to the linguistic-related activities of the translators. From the freelance translation business point of view, existing models do not consider any managerial or organisational issues (e.g. ICT and business strategy) regarding the adoption of CAT tools or other current ICT that is also part of the translators' workflow.

3.4 CAT tools and freelance translators

In this section the focus is placed on CAT tools, their role within the freelance translator workflow, and their use by freelance translators.

The discussions about the "translator's workstation" and the models reviewed in the previous section tend to be restricted to ICT to support what might be understood as "core" translation activities, such as document production, managing terminology, storing and retrieving segments of previously translated text, and automated translation. The software to support these core activities is typically categorised according to levels of automation, ranging from basic word processing facilities to support human translation, through to machine translation to support fully automated translation. CAT tools, mainly through its terminology management and translation memory functions, are present in these translator workstations, providing support to these "core translation" activities.

However, as pointed out by Varona (2002) and Locke (2005), the freelance translator's workflow involves a broader range of activities than the conventional core translation tasks. Locke cites budgeting, pricing, and hardware and software acquisition as examples of activities the freelancer must undertake (p. 20). Varona adds other activities that translation micro businesses must take on, such as file or document management, customer services, or marketing functions (p. 202).

Like the discussions in the literature about translators' workstations, existing empirical investigations of the adoption of CAT tools by translators have not tended to cover the range of activities undertaken by freelancers surrounding the core translation process. Some, for example, have concentrated on the translation activities and on a narrow subset of translation tools, such as investigations into the uptake of machine translation systems (Brace et al., 1995). Others have been devoted to the use being made of ICT within an individual organisational setting. Examples include the reviews of translation tools usage at the European Commission reported in Blatt (1998) and Brace (2000), as well as a study of terminology management tools at Ericsson (Jaekel, 2000), and a study of machine translation usage at Caterpillar (Lockwood, 2000).

Research focused on the different workflow patterns and consequences originated from the use of translation tools in translation departments of large organisations has suggested that these tools are likely to be of benefit to almost everybody, either by providing a more familiar and easy to use computer-based environment, or by suiting to specific situations and to different contexts of work (King, 1998). However, while some studies have been more comprehensive in their coverage of translators' working practices and the technology used, their findings did not really provide any detailed insights into the technology actually being used in the freelance translator community.

As highlighted in the previous sections of this chapter, CAT tools have only started to be efficient for translators from around the year 2000, mostly with the development of translation memory technology and its integration with other tools, such as terminology management systems. Evidence regarding the uptake of CAT tools by freelance translators, the benefits of using these tools, and the problems associated with their use has been reported in a few studies since 2003, which are briefly introduced here.

The eCoLoRe Translation Memory Survey 2003 (Höcker, 2003) covered a wide range of TM users to provide an appraisal of different tools aiming to investigate training issues deriving from use of TM systems. Their results were based on the responses from 208 members of German and British professional institutes of translators, the vast majority of them being freelancers, and reported a high uptake of TM (64%) with different degrees of usage, although only a third used them on a daily basis.

Fulford and Granell-Zafra, (2004; 2005) provided a broad overview of ICT used by UK translators in their study, aimed at exploring the uptake and perceptions of tools and language resources, both general-purpose and translation-specific ones, identifying the strategies employed for integrating them into the translators' workflow, and determining the impacts ICT may have on their working environments. In this case, the sample size was 439 translators, most of which (391) were freelancers and the findings revealed a widespread adoption of general-purpose software applications, but only limited uptake of more translation special-purpose software, such as TM (28%) or terminology management tools (24%), a fact that was related to a lack of awareness of, and familiarity with, these tools (almost half of the translators were not familiarised with them). In addition, this study provided insights into ICT adoption decisions linked to expectancy for improvement of efficiency and productivity and a lack of formal ICT investment strategies.

Follow-up research was carried out by Granell-Zafra, (2006) focusing on two groups of freelancers: 19 adopters and 34 non-adopters of TM and terminology management tools. The aim of this second stage of the research initiated by Fulford and Granell-Zafra was to investigate what factors were driving the adoption of these tools by means of applying Moore and Benbasat's instrument for measuring perceptions of adopting an information technology innovation (1991), as well as the impact of CAT tool adoption on translators' performance, this time by means of a performance scale based on an instrument developed by Khandwalla (1977). On the adopters' side were the perceived advantages of the tools, such as increasing productivity, enhancing effectiveness, or making the translation job easier. On the opposite, the main deterrent was found to be the perceived difficulty of learning to use the tools. In addition, impacts of adopting CAT tools reported were largely positive, and included an increase in the quality of the translations undertaken and an increased productivity.

Dillon and Fraser, (2006) reported a similar study to a part of Fulford and Granell-Zafra's research, the one focused on translators' perceptions of TM using Moore and Benbasat's instrument. Their study aimed at further understanding the perceptions of TM among individual professional translators by examining their attitudes towards TM adoption within the context of their interaction with peers and the broader industry, working environment, attitudes and motivation. The results reported by this study, based on the responses of 59 UK-based translators belonging to professional institutes and to the university of one of the researchers, showed a higher rate of TM use (52%) among most experienced practitioners, which according to the authors was influenced by the (high) level of experience of their sample. However, another conclusion of this study pointed out that overall poor levels of TM adoption were due to a lack of awareness of, and familiarity with, these tools, since there is "a large section of the translation community that does not fully understand the benefits and limitations of TM use and cannot therefore make informed decisions on the usefulness of its adoption in their working environment" (idem, p. 76), thus further confirming and validating this aspect of Fulford and Granell-Zafra's research.

Lagoudaki (2006) carried out a Translation Memory survey that covered a wider sample than previous studies (785 translators from 54 countries, most of which were freelancers) and reported a much higher use of TM (82.5%). This fact, which differed considerably from other studies, might be due to the openness of the sample of the study: "made up exclusively of those with access to the Internet" (idem, p. 27). An online survey was used to collect data to all translators, regardless of their geographical location, professional status or affiliation to professional bodies, thus, potentially attracting mainly translators who might be interested in this topic, which might explain such a high rate of TM usage. The focus was set again on gaining insights into users' perspective of TM, on reporting on the work practices related to TM tools, and on exploring the factors that affect TM use according to functional and non-functional criteria based on the EAGLES (King, 1997) and Höge's (2002) frameworks.

In summary, existing studies and research conducted in the translation sector present a range of tools that are available today to support translators in their translation activity and meet an increased demand for translation services. Among the range of translation tools available, CAT tools seem to represent a major advantage for making translators more productive and increase the quality of their translations. Although there is substantial evidence of these benefits in larger organisational contexts, research has only started to address the particular context of freelance translators in the last decade. In addition, the context of this type of translation business has not been much discussed by existing models and categorisations of the ICT available to translators.

Information Literacy and Multilingual Information Management

4

> "Tomorrow's illiterate will not be the man who can't read; he will be the man who has not learned how to learn."
>
> *(Herbert Gerjuoy in Toffler 1970, p. 414)*

4.1 Information Literacy and Multilingual Information Professionals

Information and knowledge have become valuable commodities today, being main sources of wealth (Drucker, 1993, p. 183). In fact, information is the most traded resource of the knowledge economy (Lloyd, 2011, p. 294). The information and knowledge society of the beginning of the 21st century is widely globalised and digitalised, as well as heavily characterised by great linguistic and cultural diversity. There are a number of factors that continuously increase the communication and understanding needs of multilingual environments, such as the globalisation of markets, technological advances, the consolidation of the Internet, the widespread use of electronic transactions and communications, the ever-increasing presence of digital documents, to name but a few. Information society experts at UNESCO have pointed out the essential role information competences play in education, in lifelong learning, in democracy and in human rights, and acknowledge socio-cultural diversity as a universal need (UNESCO 2006).

In this context, those professionals who facilitate communication between different languages and cultures are needed more than ever. They must be able to understand the information and knowledge that fuels an ever-changing society and be capable of adapting themselves and evolving together with this society. Language service providers, such as translators and interpreters, as well as other related services, such as cultural mediators and technical writers, face a great challenge to meet the demanding requirements of today's global market. Due to the changeable nature of the digital era, both in terms of knowledge and technology available, these professionals must continuously develop their information competences and cope with new realities (Sales and Pinto, 2011).

This challenge is closely related to meeting the information literacy standards required for higher education professionals (Bundy, 2004a): being ready to adopt and make an efficient use of current and future ICT; being capable of determining the nature and extent of the information needed; being aware of the ways to access, critically evaluate and process the information needed effectively and efficiently; and being capable of organising and reusing knowledge generated in specialised contexts to accomplish a specific purpose. The need for including IL in the Higher Education curricula has been thoroughly discussed in academic circles with a particular emphasis on its key role in the EHEA setting, which has led to developing and applying academic policies, both at local and European levels (Basili, 2008; Virkus, 2012). Being that higher education is the most suitable context in which developing IL, most research efforts have studied it from the educational perspective and applied approaches have tended to focus on the development of instrumental library and computer skills in these settings (Bawden, 2001; Virkus, 2003; Lloyd, 2010, p. 37).

In addition, the academic library has also traditionally been a common player in developing the core skills revolving around IL (Mutch, 1997; Andretta, 2005, p. 5; Gómez-Hernández, 2010) and the largest production of research about IL applied to a working setting has been that coming from Library and Information Science about librarians (Virkus 2013, p. 250). Some scholars have already noted that the discussions of librarians tend to revolve around the issues of their profession, without much emphasis on building theory or addressing the contexts beyond their particular context. Lloyd goes even further and includes IL researchers among those who are somewhat "trapped within the discursive formation of [the librarians'] profession" and states that this profession is "more focused on developing generic library skills, rather than understanding the nature of information literacy practice and the ongoing social processes that enable it" (2010, p. 181). In fact, as stated by Virkus, "during the last decade [2003–2013] our understanding of IL has sifted from skills-based approaches towards a broader and more social understanding of information practice" (Virkus, 2013, p. 255).

ICTs are undoubtedly a big ally in pursuing this goal. It is also true that the breadth and availability of technology tools and resources are wider than ever; and that higher education institutions are continuously required to make efforts to catch up with latest developments and to train people who might be able to use the latest ICT available. However, as pointed out previously, the very same constantly changing nature of the knowledge society requires a continuous learning effort and the capacity of adapting ourselves to the realities of every working environment. Technological innovation and information overload are two of the main challenges for multilingual information professionals. If they want to remain competitive, they need to make the most of their resources and find innovative solutions that enable them to deliver cost-effective quality services by not only developing their computer literacy, but also a wider information literacy competence.

4.2 Information Literacy defined

The concept of information literacy has been addressed by a considerable body of literature over the last 40 years, mostly from the domains of library and information science and computing science (Mutch, 1997), and has been connected with that of a number of literacies (Bawden, 2001), such as information, computer, IT, library, media, Internet and digital literacies. In view of such an open concept, scholars of a well-known reputation in the field like Susie Andretta have highlighted and analysed this concept as a multifaceted one (Andretta, 2005, p. 12).

The first published definition of the term IL is attributed to Paul Zurkowski, the president of the US Information Industry Association, in a report written in 1974 (Zurkowski, 1974). He defined information literates as "people trained in the application of information resources to their work [who] have learned techniques and skills for utilizing the wide range of information tools as well as primary sources in molding information solutions to their problems" (ibid. p. 6). Therefore, Zurkowski's view of IL was already focused on developing the ability to cope with the challenges of the information age within the workplace. Nevertheless, most of the efforts towards theorising about IL have been made in the context of higher education, which is undoubtedly the natural and most suitable place to introduce any training that contributes to the development of information-related competences of future educated professionals. However, as initially pointed out by Zurkowski, this development continues in the working practice environment and is influenced by the specific context of each workplace and the social processes that take place in it. This setting presents a greater level of complexity and diversity (Kuhlthau and Tama, 2001; Oman, 2001; Lloyd, 2003; Kirton and Barham, 2005) and yet, there is a scarcity of construct models or frameworks that conceptualise the practice and activities around IL in the workplace, that gain insights into settings other than the educational arena, or that investigate the effect of the collaborative nature and human interaction (Lloyd, 2010, p. 71). Thus, a detailed level of analysis is required if an understanding of IL within the workplace is pursued, since as Lloyd further states "[l]earning about the requirements and practices of work occurs at both formal and informal levels [and] requires access to both explicit and tacit sources of information. Information literacy may not follow the systematic research-based process that is advocated by the higher education setting" (ibid).

In the HE and library contexts where IL has been more intensively discussed and researched (Kirton and Barham, 2005, p. 365), the most frequently used definition is that of the American Library Association (ALA) Presidential Committee on Information Literacy: Information literacy is a set of abilities requiring individuals to "recognize when information is needed and have the ability to locate, evaluate, and use effectively the needed information" (American Library Association Presidential Committee on Information Literacy, 1989). Therefore, according to ALA, an information literate individual should be able to:

- determine the extent of information needed;
- access the needed information effectively and efficiently;
- evaluate information and its sources critically;
- incorporate selected information into one's knowledge base;
- use information effectively to accomplish a specific purpose;
- understand the economic, legal and social issues surrounding the use of information, and access and use information ethically and legally.

If, as Zurkowski already stated, IL is observed within a working setting, other elements can be taken into consideration in defining this concept. For instance, Cheuk expanded this view to involve source and end users of information to define IL as "a set of abilities for employees to recognize when information is needed and to locate, evaluate, organize and use information effectively, as well as the abilities to create, package and present information effectively to the intended audience" (Cheuk, 2002). Lloyd also broadened the scope of IL to understand it as a socio-cultural practice and process and provided the following definition:

> *"A socio-cultural practice that facilitates knowledge of information sources within an environment and an understanding of how these sources and the activities used to access them is constructed through discourse. Information literacy is constituted through the connections that exist between people, artifacts, texts and bodily experiences that enable individuals to develop both subjective and intersubjective positions. Information literacy is a way of knowing the many environments that constitute an individual in the world. It is a catalyst that informs practice and is in turn informed by it"*
>
> *(Lloyd, 2010, p. 26).*

Another important issue arises from the traditional simplification of IL to mere training or instruction in finding information or developing ICT skills to a broader perspective of information literacy that involves all aspects of information use (Mutch, 1997; Bawden, 2001, p. 225; Kirton and Barham, 2005, p. 369). In this sense, authors such as Catts and Lau (2008, p. 14) have identified the essential element that differentiates Information Literacy from ICT skills as "the transformation of information into knowledge" and the use or transmission (dissemination) of this knowledge created by the individual, as opposed to merely receiving and transmitting information using ICT. This does not mean, nevertheless, that the capacity to access and to use ICT may not include some elements of information literacy.

4.3 Information Literacy models and perspectives

The transversality of IL-related knowledge and skills, both in horizontal and vertical directions, has originated research and practice efforts at all levels of the education system. Initially, libraries were more interested in fostering information skills, such as the British Library (Marland, 1981). As a result, by the end of the 1970s, a number of training initiatives were launched with a focus on the primary and secondary education levels (Markless and Streatfield, 2007) and models like "The Big Six Skills Approach" (Eisenberg and Berkowitz, 1990) were developed to address these skills. Later on, the academic community, mostly from the Library and Information Science domain, also found this work should deserve more attention and Information Literacy was eventually developed as a discipline that addressed the information needs of more specialised educational settings, such as higher education (Bruce et al., 2000; Johnston and Webber, 2003; Andretta, 2005).

One of the consequences of this renewed interest in IL issues was the development of several theoretical frameworks to understand IL, such as the Seven Faces of Information Literacy in Higher Education (Bruce, 1997), SCONUL's Seven Pillars of Information Literacy, first presented in 1999 and subsequently revised on several occasions until 2011 (SCONUL, 2011), or the The Big Blue proposal (JISC, 2002), jointly reached by UK universities that were part of the Joint Information Systems Committee (JISC). In addition, a number of standards were developed worldwide to address IL requirements. Namely, those defined by the US Association of College and Research Libraries (ACRL, 2000) and their adaptation to the Australasian context made by the Council of Australian University Librarians (CAUL), namely the Australian and New Zealand Information Literacy (ANZIIL) framework (Bundy, 2004). A comprehensive revision of IL programs all over the world was published in the work edited by Bruce et al. (2000). In addition, Andretta dedicates a whole chapter to compare ACRL, ANZIIL and SCONUL's frameworks (Andretta, 2005, Chapter 3) and concludes that the three understood IL as the common process proposed by ALA (i.e. "involving the initial acknowledgement of the need for information, followed by competences in locating, evaluating and using that information effectively"), while the main difference rested in the linear knowledge creation process represented by SCONUL, opposed to the "recursive knowledge construction approach" of ACRL and ANZIIL.

4.4 Information Literacy in the workplace

From the context of the workplace and applied settings of IL outside the library and education, although with a much lesser intensity (Kirton and Barham, 2005, p. 365), there is also an emerging interest to study IL in the workplace that has already generated a number of investigations. Cheuk (1998) focused her research on the professional setting of auditors and provided evidence of a lack of information literacy skills, an unstructured and unpredictable information seeking process among this body of workers. Bruce's work (1999) drew on the experiences of information use in the higher education workplace and led to the development of one of the theoretical models mentioned in the previous section. Although focused on the same educational setting that the model addresses, her research involved diverse types of university staff, such as lecturers, librarians and IT professionals.

Kuhlthau and Tama (2001) conducted an exploratory study on the information seeking process of lawyers to investigate their use of sources, systems and services to accomplish their work. This research highlighted the complexity of developing tailored information systems that supported the tasks beyond the routine ones and for very particular purposes. It also showed that given the overwhelming amount of information lawyers had to work with, they needed to filter it through multilayered information systems that provided a range of functions, including organising office files, searching the Internet and handling email, and supporting the construction of cases for trial by accessing a wide range of information sources.

Kirk (2004) studied information use among senior managers from two public sector organisations in the cultural industries and identified five dimensions of IL present in this setting and that highlight the complexity of IL and the "social use of information not only in relation to information sources but also in relation to the development of new knowledge and insights" within the culture and values of the organisation (idem, p. 2). Lloyd also studied IL within a working setting as a socio-cultural practice, but outside the environment of knowledge workers. She conducted research about the information practices within the emergency services sector, namely firefighters (Lloyd, 2004) and ambulance officers (Lloyd, 2007), which allowed her to gain insights into the relationships between the roles of the workers of this community of practice and accessing, sharing and interpreting information.

In discussing how IL is seen from the business world, O'Sullivan (2002) advocates further communication and closer involvement between corporate information professionals and the managers and staff of their organisations. To do so, she provides a comparison of information literacy perspective and the business community understanding of information and knowledge management as a source of success. Although she acknowledges that IL scholars and organisations use different languages, she also highlights that "the business community is in a heightened state of awareness about the value of information and knowledge" (idem, p. 7) and reports on evidence of IL competence demand from organisations to enable effective knowledge management among their workers.

Another interesting contribution to this discussion comes from Cheuk's white paper prepared for UNESCO, the U.S. National Commission on Libraries and Information Science, and the National Forum on Information Literacy (Cheuk, 2002). In this document, she presents nine real life examples, based on companies' case studies, of a lack of IL skills in the workplace that can have a negative effect organisations' efficiency, namely:

- inability to determine the nature and the extent of the information needed;
- inability to retrieve information effectively from the information systems;
- unawareness of the full range of resources available;
- inability to evaluate and filter information;
- information and electronic mailbox overload;
- inability to exploit technology to manage information;
- inability to relate information creation and use to a broader context;
- unethical use of information;
- inability to evaluate the costs and benefits of information management.

In relation to these, she reports on several best practices that were followed by the companies under study to promote IL, although she warned that these practices were "not yet being widely adopted in business organizations" and that "most companies [were] still in the infancy stage of promoting information literacy." They included:

- further education in new technologies that enables a better understanding of how they can help achieve business goals;
- including information literacy curriculum in training and continuous professional development programs;
- increasing employees' awareness that they are knowledge workers and that accessing and using information is part of their day-to-day work;
- recognising that information literacy is a critical business skill that is as important as project management, communication skills or presentation skills;
- giving tangible rewards to employees who create quality information, who are willing to share information, and who can organise and handle information effectively.

Cheuk's cases were based on large organisations, but the issues highlighted in her white paper are also a common flaw in the small business sector (Rosenberg 2002).

4.5 Training information literate MIPs

As explained in the previous sections of this chapter, Information Literacy does not just imply a set of skills that a person must have, but is a much rather complex and dynamic practice driven by the context of the workplace (Lloyd, 2010, p. 28) and the community of practice in which it occurs (Tuominen et al., 2005, p. 341). It determines the ways in which information is generated, accessed, processed and used; and this can vary considerably from one field of knowledge or community of practice to another. "Information literacy is not always an explicit practice and is, therefore, not completely visible to outsiders who wish to research and understand" (Lloyd, 2010, p. 29). In addition to Lloyd's work (also acknowledged in Lloyd, 2006), a number of scholars have researched IL within working environments and professional communities of practice (Goad, 2002; Elmborg, 2006; Johannisson and Sundin, 2007; Hepworth and Smith, 2008; Sales and Pinto, 2011).

Focusing on higher education training from the perspective of information competences, IL didactic proposals have been best accepted when based on models and standards that guided their application (Loertscher and Woolls, 2002). Some examples are Kuhlthau's model (1991), focused on the process of searching information and on how users face the different stages of a research process; the Big Six Model (Eisenberg and Berkowitz, 1990), widely used for presenting a simple systematic framework addressed to students in order to solve information related problems in six steps; and Markless and Streatfield's model (2007), which provides an IL framework based on the processes of users' connection with, usage of, and interaction with information, driven by learning-related actions at each stage and the choice and reflection on the strategies and results of each process.

In the particular case of translator training (as well as other MIPs), scholars such as Pinto and Sales focused on the information competence acquisition within this particular community by conducting research from a user-centred perspective. They followed a holistic approach to diagnose IL needs and to develop training strategies accordingly that took into consideration three key groups of agents: translation trainees in HE (Pinto and Sales, 2007), teachers and academics (Pinto and Sales, 2008), and professionals (Sales and Pinto, 2011). In this effort, a theoretical model, InfoLiTrans (Information Literacy for Translators) was developed (Pinto and Sales, 2008) and provided outputs such as the InfoLiTrans Test. This test was devised to assess the acquisition of the information competence of translation trainees by diagnosing their proficiency in the four main areas of information search, assessment of information, information treatment, and communication and dissemination of information (Pinto et al., 2014).

IL has also been present in the theoretical models of translation competence, such as the models put forward by Kelly (2002, p. 14) and the PACTE group (2005, p. 610), where an "instrumental sub-competence" is listed among the sub-competences that are specific to the translation competence. It relates to the use of documentary resources and ICT applied to translation, and was considered as "a further characteristic of expertise in translation" after conducting additional empirical research (PACTE group, 2009, p. 227). Göpferich further validated this sub-competence as part of her translation competence model, based on PACTE's (Göpferich, 2009, p. 21) and acknowledged a general assumption about this matter among scholars (ibid., p. 12). Other scholars have termed this competence as "documentary competence" (Palomares Perraut and Pinto Molina 2000, p. 100; Gonzalo García, 2004, p. 276).

Similarly, the European Master's in Translation (EMT) expert group defined a set of competences for professional translators, as experts in multilingual and multimedia communication, that included IL-related competences such as the "information mining competence," the "thematic competence" and the "technological competence" (EMT expert group, 2009). More recently, Massey and Ehrensberger-Dow also recognised these applications of IL-related competences to the field of translation, although they did not deem it to be linked to translation competence models (Massey and Ehrensberger-Dow, 2011, p. 194). In fact, they found that there is a lack of studies addressing IL and translation when they state that "[t]he relative weight accorded to information literacy by translation practitioners, teachers, and scholars has yet to be underpinned by a significant body of research" and that "[t]he questions of how and whether translators use the far wider range of electronic and non-electronic resources, both linguistic and extra-linguistic, now at their disposal [...], and of how novices and professionals differ in this regard, remain to be investigated in detail" (idem., p. 195). This claim for further empirical research motivated their study *Translation Tools in the Workplace*, aimed at exploring ICT and research tools and resources' impact on professional translation processes, and took into consideration some previous studies, such as Fulford (2001), Nord (2002) and Pinto and Sales (2008), among others, although they did not refer to research reported in Reuther (1999), Fraser and Gold (2000), Fulford (2002b), Höcker (2003), Fulford and Granell-Zafra (2003; 2004; 2005; 2008), Lagoudaki (2006), Dillon and Fraser (2006), or Alcina et al. (2007).

In addition to these initiatives, there are more studies in progress that are trying to investigate how IL affects translation from an information behaviour angle, such as Enríquez-Raído's (2011) study on developing web searching skills in translator training, or an ongoing research work conducted by Sales, Pinto and Granell about the information search process in documentary tasks when translating (Forthcoming). It is definitely a must to foster and conduct more empirical research about IL and translators to inform and improve their training, although it may also be necessary to adopt an integrated view of information literacy for translators, whether it is related to translation competence models or not, that builds upon the vast body of IL literature, as already done by Pinto and Sales, and that puts together all previous efforts to research IL-related issues in the translation context.

As a result of this growing interest in improving IL among the translation community and applying the training proposals arising from research, several efforts to transfer research contributions to real training practice have already seen daylight. Some initiatives are focused on fostering and improving training provided at HE institutions and some others also try to provide a continuous training assistance in consonance with the IL paradigm and leading towards enabling a successful lifelong learning. Among the former, there are many universities using student-centred methodologies to develop information competences in their translation degrees, either through specific courses on documentary research, as a transversal competence present in ICT-related subjects (i.e. courses on IT applied to translation, terminology, and resources for translators) or most subjects or modules devoted to translation in specialised domains (OPTIMALE 2012).

In Spain, for example, "Documentary Research Applied to Translation" *(Documentación aplicada a la traducción)* has been a compulsory core course of Translation degrees' syllabi since these HE studies were initiated in 1991, a need that was further highlighted in the development of new degrees under the EHEA framework (ANECA, 2004). Several specialised translation master degrees also offer specific programmes including information-related skills, required for all students regardless of the specialisation, as pointed out by the OPTIMALE Academic Network (Optimising Professional Translator Training in a Multilingual Europe), which involves 70 partners from 32 different European countries (OPTIMALE 2012). One example of the latter is the 6 ECTS credit subject "Professional Practice, Medical Terminology and Medical Information Sources" *(Práctica professional, terminología y fuentes de información)* at the Master's Degree in Medical and Healthcare Translation Master at Universitat Jaume I (Department of Translation and Communication at the Universitat Jaume I 2013).

In addition to this formal presence in HE syllabi, some didactic proposals for improving IL among translators are also trying to bridge the gap between translators and information at different levels of expertise: early undergraduate trainees, undergraduate trainees in subsequent courses during their translation studies, students of master courses on specialised translation/interpreting/cultural mediation, or throughout MIPs' professional life. A good example of this is ALFINTRA[1] (Information Literacy for Translators, from the Spanish name *Alfabetización Informacional para Traductores*), a didactic proposal that applies the InfoLiTrans theoretical model, developed by Pinto and Sales (2008). It is a multifaceted approach that puts together a set of elements (namely, knowledge, ICT, resources and processes) with competences and skills to provide an e-learning portal based on web resources. The design of ALFINTRA drew on the results obtained from the InfoLiTrans project, in which a number of competences were deemed to be reinforced (namely, information management, organisation and planning, and ICT-related skills) to measure the acquisition of information competences and aptitudes by students, but also to diagnose them and to solve the problems that have been detected in the learning process. The resulting tool is an autonomous e-learning platform that includes information literacy fundamental knowledge addressed to the translation community and is structured according to the issues around information management needs, including contents, activities, FAQ-like sections, multimedia materials, diagrams, and concept maps, in addition to a selection of resources and complementary bibliography.

Another contribution of ALFINTRA, resulting from the expert input provided by translation and interpreting lecturers during the InfoLiTrans project, is its compilation of special-purpose resources aimed at addressing the needs of the different fields of specialised translation and interpreting (i.e. legal translation, scientific and technical translation, medical translation, audiovisual translation, localisation, literary translation), apart from those of general-purpose translation and of translation-related research.

---

[1] This portal can be accessed at *http://www.mariapinto.es/alfintra/*

Either during formal training at HE institutions or by lifelong learning initiatives that foster a continuous professional development, what seems clear is that each person must be able to develop his or her information competences for critical thinking, decision-making and problem-solving purposes in order to be effective and remain competitive in the global arena we live in. In this ever learning environment, the role of tutors and more experienced colleagues needs to shift from the "sage on the stage [to] the guide on the side," as already highlighted by King in the early 1990s (1993, p. 30) and put forward by the current EHEA.

An apt conclusion to this chapter is the following quote from O'Hagan and Ashworth about their prediction of the translators' environment in 2002 that seems to be happening today more than ever before: "Understanding the nature of the change that is taking place all around translators and interpreters will better prepare them to face further challenges" (O'Hagan and Ashworth, 2002, p. 130).

A strategic approach to adopt ICT: from using information and communication technology to making use of information and technology to communicate

This chapter focuses on discussing how an information systems approach can help MIPs make a strategic use of ICT to achieve their business goals and be productive while optimising efficiency and effectiveness. First, an introduction to a strategic approach is presented, then the concept of information systems strategy and what it involves is defined, and finally a review of relevant literature from the bodies of small business management and information systems research is presented to provide a solid basis that can be used to better understand IS and ICT adoption in the setting up of translation businesses.

5.1 The Information Systems approach to ICT

As explained in Chapter 2, the information systems of an organisation consist of the information technology infrastructure, data, application systems and personnel that employ ICT to deliver information and communications services in an organisation, but it also refers to the management of the organisational function in charge of planning, designing, developing, implementing and operating the systems and providing services (Davis, 2000). Thus, the concept of IS combines both the technical components and the human activities within an organisation and describes the process of managing the life cycle of organisational practices. The main goal of information systems is to gain a competitive advantage through efficient and effective use of the human and technology resources of an organisation and multilingual information professionals, either as freelancers or as part of small or micro businesses. These businesses are in great need of getting things done right in their day-to-day activities (i.e. being efficient) and of doing the right thing to provide quality solutions that meet the requirements of the market (i.e. optimising their effectiveness to gain a competitive advantage). Information systems have already been going down this path for a long time; however, MIPs still have a long way to go in this sense.

It is not a matter of how many tools you might master or how many resources you might gather and have within your reach; it is not a matter of how much you know about a particular domain of knowledge or how much you invest in training or acquiring the latest technology. It is about how well you are capable of deciding which are the best tools or technology you need for your particular working setting; it is about being capable of organising your resources effectively to retrieve the very particular piece of information you need in the shortest time to remain productive and deliver a consistent and quality service; it is about being capable of recognising your information needs at every moment and learning to evolve and adapt your business to the changes and new developments; and it is about developing a knowledge-based system that is tailored to your business needs and that involves all the human and material resources that are part of your working setting. All in all, it is about adopting a strategic view that is aimed at being aware of your working setting and making the right decisions to be efficient and effective.

It is clear that IS were created to address the needs of large, complex organisations, with large scale and scope IS, and best fit this type of organisation. It might seem that a much simpler organisation, such as an SME or a micro business, is in no such need to use heavy-duty decision-making machinery, but still there are many advantages that can be adapted to the particular context of smaller organisations to successfully undertake an information-related activity (Yap et al., 1992; Thong, 1999; Tapscott, 2004). Given the characteristics of MIPs' business, and that this type of activity deals intensively with information and knowledge, it makes them likely candidates to benefit from IS.

Given the lack of formal approaches to address the organisational complexity of translation businesses and the lack of formal frameworks to drive ICT-related decisions and processes in this setting, the lessons learned from the field of business strategy and information systems can provide a useful reference to underpin an IS strategy for MIPs and to reach the strategic fit that enables productivity and quality services. A freelance business requires a much simplified architecture than a large organisation, which on one hand reduces the number of functions and systems to be coordinated, but on the other concentrates the different functions of the business into one person, i.e. the manager, thus increasing the volume of managerial and administrative tasks that affect the core activity and the degree of dependence on this person.

IS are aimed at empowering managers, engineers, and ICT users with knowledge and techniques for effective decision making. In the case of multilingual information professional practice in a freelance context, the same person needs to assume all these roles in with a greater or lesser level of intensity, depending on the type of projects undertaken. It is clear that the theory and logic for managing an organisation involves a considerably more complex scheme than in the case of micro SMEs like freelancers. It is also clear that information systems support the objectives of organisations and their rationality by providing support to analytical processes. Then, why should freelancers design an information system aimed at formalising the use of ICT and the management of the business? Why even just think about an ideal framework if their (micro) organisation is much simpler? The reason is that, even though organisations never function according to the ideal – neither do the large ones (Davis, 2000, p. 65) – using information systems adds coherence to deploying ICT and to decision making.

Therefore, an IS strategy approach can help MIPs to apply ICT appropriately in a timely way and in harmony with their business strategies, goals and needs.

5.2 Information Systems strategy

IS strategy is a complex concept which includes three streams of literature (Chen et al., 2010): strategic information systems planning (SISP) (Galliers, 1991; Premkumar and King, 1994; Peppard and Ward, 2004), alignment between IS strategy and business strategy (Henderson and Venkatraman, 1993; Chan et al., 1997; Chan and Reich, 2007), and competitive use of IS or using IS for competitive advantage (Melville et al., 2004; Wade and Hulland, 2004; Piccoli and Ives, 2005).

Chen et al. define IS strategy as "an organizational perspective on the investment in, deployment, use, and management of information systems" (2010, p. 235) in their comprehensive review and analysis of this concept and differentiate three conceptions of IS Strategy (2010, p. 238), namely, "(1) IS strategy as the use of IS to support business strategy; (2) IS strategy as the master plan of the IS function; and (3) IS strategy as the shared view of the IS role within the organization."

In the particular context of multilingual information professionals, the first conception, IS strategy as the use of IS to support business strategy, assumes that a particular business strategy is already guiding the professional activity, which does not usually seem to be the case for translators, or at least a formally defined one (Granell-Zafra, 2006, p. 215). Still, if this strategic view is to be adapted to this setting it is important to outline the objectives pursued by the translation business in the short, medium and long term and make them explicit to better assess how an IS strategy can help gaining or sustaining the targeted competitive advantage. This is usually called "alignment" or "fit" between the business strategy and the IS strategy, and, although not in managerial terms, it has also been acknowledged in the translation sector by authors like King, who outlines as a critical factor the need for a preliminary analysis before adopting translation tools and states that "maximum benefit from introducing translation technology can be gained by careful preliminary analysis of what is really needed and of the consequences of introducing it" (King, 1998). This strategic approach to define the IS strategy of a MIP's business helps focus on core customers to improve effectiveness, defining a suitable organisational structure that is aligned with the objective of the business, and having a long-term thinking capable of achieving future goals.

Chen et al.'s second conception of IS strategy, i.e. IS strategy as the master plan of the IS function, is more specific to the IS function within the business and focuses on defining the particular IS assets (information needs, hardware, software, communications, data, people, training) required to reach the desired level of efficiency as well as on planning how to structure them to best accomplish day-to-day operations. It can be applied to organisations that do not have a clear business strategy or that look for functionality over gaining a competitive advantage (Duncan, 1995; Byrd and Turner, 2001; Bhatt et al., 2005). This tactical approach to defining the IS strategy of a MIP's business can enhance the visibility of the work processes and practices to ensure that work is being done towards fulfilling the objectives of the business (i.e. to clearly know why and how things are being done) and to improve operational and communicative practices dynamically to adapt the business to possible changes in the environment.

The third conception of IS strategy, i.e. IS strategy as the shared view of the IS role within the organisation, has to do with the overall perspective towards IS. It is necessary, from a strategic point of view, to share a common vision across the organisation that guides any ICT or training investments, as well as deployment decisions. In the context of MIPs' businesses, which might not involve many people, it is still very important to position the identity of the organisation from a cultural, technological, social and business viewpoint. This can not only be helpful in guiding any innovations and investments internally, but also in engaging colleagues and collaborators, and in involving customers and providers into the culture of the business services provided.

IS are not only developed because organisations (or their managers) want to do things faster or because they want to have the latest and greatest technology. IS are developed strategically to help gain or sustain some competitive advantage over rivals (Porter and Millar, 1985). Although the approach followed in this book is that of a broader conception of an IS (i.e. that involving the three conceptions of IS strategy detailed above), some scholars differentiate functional IS from those IS particularly aimed at gaining a competitive advantage, called "Strategic IS."

Porter (1979; 2008) identifies five competitive forces that operate in a business environment and that can help each organisation to better address the risks posed by its market:

- the threat of new entrants;
- the bargaining power of suppliers;
- the bargaining power of customers;
- the threat of substitute products or services;
- the rivalry among existing competitors.

Technology use can be a powerful enabler of competitive advantage and for achieving innovative ways of developing multilingual communication effectively, thus differentiating the services provided by MIPs from the crowd of professionals competing in the global market. Therefore, IS, in addition to the IL perspective discussed in Chapter 4, can be a valuable help towards an informed approach for adopting ICT. The next section of the chapter is focused on addressing these aspects.

In summary, the objective of the IS strategy should then be aimed at defining the structure within which information, information systems and information and communication technology is to be applied within the organisation over time. An IS requires the specification of the information needs; the processes necessary to collect, produce, store and disseminate information; the systems needed to support the organisational activity in these processes; the hardware, software, communications and data facilities required; the people involved in the processes; and the competences needed to support the information systems. In order to achieve a successful IS strategy, planning is required to develop all the aspects around the strategy, to define how they should be structured, and to establish short, medium and long-term objectives and infrastructure that will allow information systems to be designed and implemented efficiently and effectively.

5.3 IS and ICT adoption in small businesses

Trying to strategically address the adoption of IS and ICT is not an easy task. Nevertheless, although CAT tool adoption has not been much researched in the translation sector, there are other areas in which research about the adoption of technologies in small businesses has been studied more extensively, mostly in the last two decades of the 20th century. This section presents a discussion of the seminal literature identified in the domains of Information Systems (IS) and small business management.

Within each of these domains, a number of more specific areas were deemed important to the present research, namely,

- Small business management:
  - IS/ICT adoption strategies in small businesses;
  - IS/ICT adoption factors in small businesses, including motivators and inhibitors;
  - the influence and role of the CEO in IS/ICT adoption decisions.
- Information systems:
  - measures for determining the success of IS/ICT adoption and implementation in organisations;
  - stage models of IS/ICT adoption in organisations.

5.3.1 ICT and SMEs

The dawn of the 21st century was marked by an information-based economy that made organisations more reliant upon Information and Communication Technology and Information Systems to support their business processes (Irani and Love, 2001a). However, research undertaken by Kempis and Ringbeck, (1999) claims that a higher availability of ICT does not always translate into higher efficiency and effectiveness, and suggests that a significant proportion of organisations may be under-performing with regard to efficiency and effectiveness of ICT utilisation. Researchers and practitioners are seeking an explanation for this fact. For example, McKay and Marshall (2001) state that the notion of an information-based economy and the arrival of an e-business domain have led to considerable faith being placed in IT to deliver performance improvements, and that there is a concern that ICT/IS is not delivering what it promises. Irani and Love (2001a; 2001b) attribute this lack of delivery to the difficulty in determining business value from ICT/IS investments, and the considerable indirect costs associated with enterprise-wide systems. The measurement of business value of ICT/IS investments has been widely debated in the IS and business management literature (see, for example, Weill and Olson, 1989; Serafeimidis and Smithson, 1996; Irani et al., 2001), yet there has been a lack of consensus in defining and measuring ICT/IS investments (Irani and Love, 2002). This area of research has looked at these issues in the broad organisational context, but it represents a more critical problem in the case of the SMEs, where management functions and ICT budgets are more limited.

In SMEs, managers play a decisive role when deciding about investing in new technologies (Cragg and King, 1993), therefore, they have to carefully consider the potential impacts of acquiring them, and then take an informed decision of the investment. In order to better utilise resources, managers need to have an understanding of the impact of ICT/IS on the organisational infrastructure and overall performance, as shown in the discussion of ICT/IS evaluation above. This literature shows that an analysis of potential impacts of ICT/IS for SMEs is needed, and once an ICT/IS has been adopted, ICT/IS evaluation would provide feedback that allows to better establish benchmarks of what is to be achieved by ICT/IS investments.

Studies on the evolution of IS in SMEs (see, for example, Saarinen, 1989; Cragg and Zinatelli, 1995) have identified several approaches to investigate the evolution of IS in organisations, although not necessarily small firms. First, a number of models have looked at the growth stages undergone by organisations adopting ICT/IS. Second, a number of factors that influence the decision of adopting ICT/IS in SMEs have also been studied to understand what are the determinants of the adoption. Other studies have looked at the factors that determine the IS success. Finally, IS evolution has also been studied through the concept of its sophistication in organisations. These four aspects of the IS literature on ICT adoption in SMEs are analysed in the following sections and related to the specific case of freelance translation businesses.

5.3.2 Models of ICT adoption in SMEs

According to Cragg and Zinatelli (1995), one of the related areas identified by researchers as relevant to understand the adoption of new ICT is the analysis of "stage models" of IS adoption and evolution in organisations, which are based on the assumption that computing moves through a series of growth stages.

Saarinen (1989) reviewed the existing literature about the evolution of an organisation's information systems through the discussion of models developed in IS science and the broader theoretical features to which they apply. This review goes from, according to King and Kraemer (1984), the important initial step in research into the evolution of IS in organisations (Churchill et al., 1969; Nolan, 1973; Nolan, 1979) to more specific-area models (IBM, 1981; McFarlan et al., 1983; Rockart, 1983; Zmud et al., 1987). The early stages of growth in these models seem to be close to each other in the models, but as the growth process proceeds, more differences in the assumed development patterns can be detected. According to Saarinen, these models explicitly or implicitly incorporate underlying theoretical principles from economics, diffusion theory, organisational learning, and growth and stages theory.

Economic theories offer a whole body of literature which could be applied in analysing the development of computing in organisations, and Saarinen cites Schumpeter (1934) as an example. These theories assume that the balance between supply and demand for ICT is reached in the same way as in the market (i.e. the price of using ICT determines the demand). However, researchers of ICT/IS evolution have considered that economic theories did not provide an answer to ICT/IS evolution, which changes over time, and that other, more descriptive, models and theories were needed to explain the mechanisms beyond the adjustment process of demand and supply (Saarinen, 1989, p. 393).

Diffusion theories define "diffusion" as "the process by which an innovation is communicated through certain channels over time among the members of a social system" (Rogers, 1995, p. 5). Saarinen (1989, p. 393) states that "diffusion theories could offer a wide and well-formulated set of models [...] meant for use in studying a phenomenon which represents one possible view of the development of computing in organisations," and that "authors in the IS field seem to have been aware of this literature, but the existing diffusion models have not been used significantly until now."

Organisational learning theory describes the changes associated with the ICT/IS evolution process through the concept of "learning curves," which illustrate, for example, unit costs as a function of the number of times performed (Saarinen, 1989, p. 394). According to Saarinen's work, "most of the IS evolution models have recognized learning to be one of the most important mechanisms. However, connections with existing learning theories seem to be weak and their potential to be only partially utilized" (idem, p. 394).

Finally, growth and stages theory describes the growth of an organisation in terms of sequences of distinguishable stages. According to this theory, organisations go from one stage to the following after reaching a crisis and undergoing a revolution that leads them to a new growth process.

Churchill's model (Churchill et al., 1969), in which the idea of the stages theory was introduced to computing, proposed a number of levels of automation, from the simplest tasks to the automation of more complex tasks (such as making decisions based on strategic purposes). The early approaches of ICT evolution in translation firms suggested in the translation tools literature seem to have followed a similar path to the stages described by Churchill. These approaches (see, for example, Hutchins and Somers, 1992; Hutchins, 1996) understood the automation of the translation process as the "logical" aim of the translation tool development and use.

Nolan's model (Nolan, 1973; Nolan, 1979) presented a more detailed account of the stages of IS growth in organisations, using budget growth as the primary indicator of the evolution, and implicitly based on the dynamic diffusion theory and organisational growth models (according to Saarinen, 1989). This model has been considered to be the most inclusive description of the evolution of IS in organisations (Saarinen, 1989), and although its validity has been criticised (Benbasat et al., 1984; King and Kraemer, 1984), researchers suggested its testing in small firms (Cooley et al., 1987; Stair et al., 1989), and studies on ICT adoption have drawn on it (see, for example, Cragg and King, 1993).

Churchill's and Nolan's models of stages of adoption have explained IS evolution in the broad context of organisations; however, many of the processes and management issues of large organisations are much simpler in the case of translation freelance businesses (see, for example, Joscelyne, 2003).

Research has showed that the models developed in the IS field do not take full advantage of the possibilities offered by the theories and models developed in the more mature fields of scientific inquiry, and that computer use is generally used to indicate the state and growth of computing, in spite of showing evidence that the extent of use (often measured by cost) has any direct effect on the level of benefits gained. Saarinen criticises that these models are descriptive and that they give no suggestions for evaluating the effectiveness of different ways of using computers, to conclude that "as long as technologies continue to develop, there will be a need for detailed models addressing the specific problems of each technology" (1989, p. 397).

In the translation sector, there have not been any models that have tried to explain ICT adoption until some have recently started to investigate certain aspects of using technology tools (cf. Chapter 3.4). For this reason, one of the aims of the present work is to look at how ICT adoption has been investigated in other disciplines, such as information systems, and use the suggestions and theoretical foundations offered by them to develop a suitable framework to investigate CAT tool adoption in the specific context of freelance translation businesses.

5.3.3 ICT adoption factors in SMEs: motivators and inhibitors

Another of the IS-related areas identified by researchers as key to understanding the adoption of new ICT is the analysis of factors that may affect the decision and the process of adoption (Cragg and Zinatelli, 1995).

In order to get an understanding of whether ICT are successfully adopted and used in firms or not, researchers have tried to identify the factors that affected positively and negatively the processes of adoption. Prior to the main stage of ICT adoption, there are a number of factors that can encourage or discourage the decision of adopting ICT, therefore leading SMEs to adoption or deterring them from adopting ICTs.

There is a body of literature that relates to ICT adoption and to the factors that encourage and discourage it. Researchers in this area which have analysed ICT adoption success factors, covered in the next section, have drawn on the literature that explores the motivators and the inhibitors for ICT growth in SMEs (for example, Cragg and King, 1993); and the literature on reasons for computerisation in SMEs (Easton et al., 1982; Farhoomand and Hrycyk, 1985; Malone, 1985; Baker, 1987; Lefebvre and Lefebvre, 1988; King and McAulay, 1989).

5.3.3.1 Motivators

Studies on ICT growth in SMEs, such as that by Cragg and King (1993), are based on previous research on factors that encourage or discourage computerisation in SMEs, and have distinguished a number of motivators that reflected internal, external and individual factors, such as relative advantage in information processing, relative advantage in planning and control, and relative advantage in work improvement. This group of motivators focuses on factors that give some kind of advantage (e.g. time, effort or economic savings), and the authors identify three more general factors related to external support (consultant support), competitors (competitive pressure) and CEO involvement (managerial enthusiasm). CEO enthusiasm toward computing was found to be the strongest motivating factor for ICT adoption and growth by Cragg and King. However, the nature of this involvement can vary from one CEO to another, as shown by Martin (1989), who revealed five types of involvement, ranging from remote to close involvement, as discussed later in this section.

Among the findings of the studies on the motivators for the computerisation of SMEs as stated above, there are some factors that have been identified as highly significant drivers to ICT adoption: the search for an increase in office task productivity (Easton et al., 1982; Baker, 1987), the improvement of information management and processing (Easton et al., 1982; Farhoomand and Hrycyk, 1985; Malone, 1985; Baker, 1987; Lefebvre and Lefebvre, 1988), and the effects of external information sources (Lefebvre and Lefebvre, 1988; King and McAulay, 1989).

More specifically, a desire for an increase in productivity, identified as a key persuasion factor by Baker, (1987), is a perceived benefit that allows SMEs to be more efficient and save time and effort (Cragg and King, 1993) through the automation of office tasks.

A higher capability of data processing (Easton et al., 1982; Baker, 1987), quicker processing of information (Lefebvre and Lefebvre, 1988), and therefore an improvement in information management are other factors that bring more savings in terms of time and effort (Cragg and King, 1993), help the firm to cope with information overload (Farhoomand and Hrycyk, 1985), and increase the performance of the firm through higher control for effective management (Malone, 1985).

External factors may also influence the decision of adopting ICT in a variety of forms; for example, through the influence of consultants that increase the willingness of CEOs to use ICT, either by recommending the firm to develop an ICT solution or by the consultant's own use of technology (King and McAulay, 1989). Lefebvre and Lefebvre (1988) found some more external sources of information affecting ICT adoption, apart from consultants' support, such as general environment, clients, competitors' pressure (also identified by Cragg and King, 1993), employees and suppliers, although they did not find that these factors are clearly more important than others for their adoption.

Lefebvre and Lefebvre (1988) discuss that the decision of the small-firm manager is mainly influenced by information sources that are external to the firm, being that the manager is the person more likely to take the final decision, especially in SMEs, where the manager is much more prone to outside influence than managers of large firms (Malone, 1985).

A summary of the factors positively affecting IS adoption in SMEs found in the reviewed literature is presented in Table 5.1. This categorisation classifies motivators into external and internal.

In the freelance context, the role of the CEO is particularly important, since the manager of the freelance translation business is also the end-user of CAT tools. In the IS literature reviewed, CEO involvement and eagerness towards technology – called "managerial enthusiasm" by Cragg and King (1993) – has been confirmed to be one of the most important factors during the stage when the adoption decision is taken and once ICT have been adopted, for success in the use of the systems (DeLone, 1988). Freelance translators have to both make the decision to adopt CAT tools and use them. Therefore, it is relevant to examine studies that have analysed CEO involvement in more depth, such as Martin (1989), who identified a range of different involvement patterns among CEOs in SMEs, and categorised them into five groups of behaviour patterns. Table 5.2 shows the five types of CEO involvement identified by Martin.

Although Martin's classification of CEO involvement looks at the role of managers in larger organisational contexts, it is important to observe that all the behaviour patterns described in levels 2 to 5 are present in the figure of the freelance translator adopting CAT tools. Level 1 cannot apply to the freelance translation context because it describes a remote involvement, which in the case of freelance translators does not exist.

The role of the CEO in freelance translation businesses, among other motivators identified (e.g. advantages of adopting ICT, influence of external sources), is something that will need to be investigated in this study to gain a better understanding of the characteristics of the freelance translators underpinning the adoption of CAT tools.

Table 5.1 Motivators affecting IS adoption in SMEs

Type	Motivator	Studies
Internal	a. Advantage in information processing	Baker, 1987 Cragg and King, 1993 Easton, 1982 Farhoomand and Hrycyk, 1985 Lefebvre and Lefebvre, 1988 Malone, 1985
	b. Advantage in planning and control	Cragg and King, 1993 Malone, 1985
	c. Increase of productivity	Baker, 1987 Cragg and King, 1993 Easton, 1982
	d. CEO involvement	Cragg and King, 1993
	e. Advantage in managing information	Cragg and King, 1993 Farhoomand and Hrycyk, 1985
External	a. Consultant support	Cragg and King, 1993 King and McAulay, 1989
	b. Competitive pressure	Cragg and King, 1993 Lefebvre and Lefebvre, 1988
	c. External information sources	King and McAulay, 1989 Lefebvre and Lefebvre, 1988
	d. Clients, employees or suppliers influence	Lefebvre and Lefebvre, 1988

Table 5.2 Types of CEO involvement in computerisation in SMEs (Martin, 1989, p. 192)

Type	Behaviour pattern
1	Top manager is remote from the computer resource, and is uninvolved even in key decisions in relation to its development or operation.
2	Top manager is involved in a managerial, supervisory capacity, and identifies goals and sets targets.
3	Top manager is closely involved in implementation, and takes part in detailed choice and/or design decisions.
4	Top manager is directly involved technically, and takes part in programming or spreadsheet development.
5	Top manager routinely interacts directly, hands-on, with the IS.

5.3.3.2 Inhibitors

In the same research conducted by Cragg and King (1993) about motivators, the inhibitors of ICT growth are also explored, again based on previous studies on computerisation in SMEs (Bourner et al., 1983; Baker, 1987; King and McAulay, 1989). According to Cragg and King's (1993) study, the most significant factors that deter

SMEs from adopting ICT identified by the authors can fall into broader categories: ICT education factors (such as lack of CEO or personnel with broad ICT knowledge, lack of personnel with specific ICT skills, and negative influence of higher levels), lack of managerial time, economic factors (such as an inappropriate economic climate, excessive costs, and firms being too small), and technical factors (such as having an unstructured system, and having poor software support).

As corroborated by other studies (Baker, 1987; Lefebvre and Lefebvre, 1988; King and McAulay, 1989), the lack of general ICT knowledge has been found to be the most important inhibitor for ICT adoption and growth, together with a lack of economic resources. These two factors become accentuated in the case of SMEs, where economic resources available for ICT investments are more limited than in large companies. The fact that many SMEs do not even have a department devoted to ICT support, together with the tendency to employ more generalist than specialist staff, makes it more difficult for them to have a high internal ICT knowledge. For this reason, another major inhibitor of ICT adoption is the influence of the person that has to make decisions regarding ICT, generally the CEO. Since CEOs may not have a high ICT knowledge either, their decision tends to depend on their enthusiasm towards technology and the confidence they may have in external support, such as consultants or vendors (Kole, 1983; Baker, 1987; King and McAulay, 1989; Gable, 1991).

A summary of the factors that deter the adoption of ICT in SMEs found in the literature is presented in Table 5.3. This categorisation classifies inhibitors into external or internal factors to the organisation.

Table 5.3 **Inhibitors affecting IS adoption in SMEs**

Type	Inhibitor	Studies
Internal	a. Lack of general IS/ICT knowledge	Baker, 1987 Cragg and King, 1993 Farhoomand and Hrycyk, 1985 King and McAulay, 1989 Lefebvre and Lefebvre, 1988
	b. Lack of managerial time	Baker, 1987 Cragg and King, 1993 King and McAulay, 1989 Lefebvre and Lefebvre, 1988
	c. Lack of economic resources	Cragg and King, 1993
	d. Inappropriate economic climate	Cragg and King, 1993
	e. Too small firm	Cragg and King, 1993
	f. Unstructured system	Cragg and King, 1993 Lefebvre and Lefebvre, 1988
External	a. Lack of good technical support	Cragg and King, 1993 Farhoomand and Hrycyk, 1985
	b. Lack of confidence in vendors	Baker, 1987 King and McAulay, 1989
	c. Technology changing quickly	Baker, 1987

The inhibitors identified in the literature about ICT and SMEs represent a valuable set of factors that need to be investigated to gain a better understanding of the characteristics of the freelance translators hindering the adoption of CAT tools. For example, factors such as the "lack of managerial time," the "lack of expertise" or the "lack of economic resources" which affect SMEs in their adoption of ICT, are particularly likely to affect micro businesses, like freelance translation businesses, where the translator is also the owner-manager of the business. Similarly, some external inhibitors that might affect freelance translators are the existence and reliability of external support in the shape of consultants and mostly vendors, who can provide freelance translators with the right training and technical support that allow them to cope with the fast changes that the technology used by freelance translators is experiencing (Pérez, 2002; Joscelyne, 2003).

5.3.4 Success factors for ICT implementation in SMEs

Success factors have also been studied in the area of information systems that looks at ICT adoption in SMEs. The concept of "information systems success," also called by some authors "IS effectiveness" (Hamilton and Chervany, 1981; Raymond, 1990; Thong et al., 1996), is recognised by many researchers as difficult to define (as shown, for example, by Weill and Baroudi, 1990).

A number of studies have investigated the factors contributing to IS success in the context of small firms. For example, Raymond (1985) investigated the relationships between organisational characteristics and IS success based on the studies of Ein-Dor and Segev (1982), DeLone (1981) and Turner (1982). Raymond used user information satisfaction and level of system satisfaction as measures of IS success, and the findings revealed that systems success was higher where a greater proportion of applications were developed and used internally, a greater number of administrative applications were used, interactive applications had been implemented, and the IS function was situated at a high organisational level.

Similarly, other studies examining the factors that affect the successful use of IS by SMEs found not only a positive association of IS success with the CEO knowledge of computers, but that CEO involvement was a key factor for IS success (DeLone, 1988; Montazemi, 1988; Palvia et al., 1994; Caldeira and Ward, 2002).

The factors affecting IS success in small businesses that were found to be significant in previous studies were categorised into four major classes (organisational characteristics, organisational action, system characteristics and internal expertise), plus a fifth category regarding external expertise by Yap et al. (1992), leading to the development of a descriptive model of key factors to IS success in a small business context.

External factors affecting IS success were further investigated by Soh et al. (1992), Palvia, (1996), Thong et al. (1996) and Igbaria et al. (1998), and the computerisation success of SMEs was associated with the capability, experience and effectiveness of the consultant.

There are some inconsistencies in the findings of all these studies; however, the positive association of IS success and a higher involvement of the CEO in SME computerisation was supported by most of the studies (DeLone, 1988; Yap et al., 1992; Palvia et al., 1994). Cragg and Zinatelli (1995) attribute this inconsistency in the findings to the evolution of IS and changes in the factors affecting its success over time.

This review of the studies carried out in the IS domain regarding IS success factors is particularly important for the present research due to the lack of much formal research focused on CAT tool adoption and the factors that contribute to its success in the translation studies area. However, the analysis of IS success factors in previous research on IS presents a number of factors (such as CEO involvement in CAT tool adoption, or the influence of software consultants/vendors) that needed to be investigated with regard to CAT tool adoption by freelance translation businesses.

A summary of the factors affecting IS success in SMEs found in the literature reviewed is presented in Table 5.4. This categorisation is based on Yap et al.'s (1992) classification, and includes a fifth category regarding external expertise identified in the same study, as well as adding factors found in subsequent studies.

5.3.5 SMEs and ICT sophistication

It has been noted in a number of studies that one of the fundamental problems that IS researchers face is to characterise organisational information systems, and particularly identify different criteria of systems "maturity" or "sophistication" (Benbasat et al., 1980; Cheney and Dickson, 1982; Ein-Dor and Segev, 1982; Saunders and Keller, 1983; Gremillion, 1984; Lehman, 1985; Mahmood and Becker, 1985; Raymond, 1988; Raymond and Paré, 1992).

"ICT sophistication" is defined by Raymond and Paré (1992) as a multi-dimensional construct which refers to the nature, complexity and interdependence of ICT usage and management in an organisation. Therefore, the concept of ICT sophistication integrates both aspects related to IS usage and IS management, also present in Nolan's model of stages of growth (Nolan, 1973; Nolan, 1979). Raymond and Paré (1992), based on variables from previous research to characterise each dimension, identified four dimensions within the construct related to technological support (technological sophistication), information content (informational sophistication), functional support (functional sophistication) and management practices (managerial sophistication).

Technological sophistication refers to the number and diversity of information technologies used by SMEs as well as to the nature of the hardware and the development tools used by the firm.

Informational sophistication refers to the nature of the application portfolio of the SME, including both transactional and administrative applications. Another aspect of informational sophistication identified by Ein-Dor and Segev (1982) relates to the degree of integration of the applications, in an SME basically characterised by the presence of software (e.g. database) or hardware (e.g. local area network) that allow information interchange and resource sharing.

Functional sophistication relates both to the structural aspects of the IS function in the SME (e.g. the location and autonomy of the IS function and the number of internal IS specialists) and to the ICT implementation process (e.g. method, source and uniqueness of applications).

Table 5.4 **Classes of factors affecting IS success in SMEs**

Class	Factors	Studies
1. Organisational characteristics	a. CBIS experience	Raymond, 1985 Igbaria et al., 1998
	b. Proportion of applications developed internally	Raymond, 1985
	c. Presence of in-house processing	Raymond, 1985
	d. User attitudes	Caldeira and Ward, 2002
	e. Rank of computer function	Raymond, 1985
	f. Financial resources	Caldeira and Ward, 2002
	g. Degree of decentralisation	Montazemi, 1988
	h. Company size	Palvia et al., 1994
	i. Age of company	Palvia et al., 1994
2. Organisational action	a. CEO support and attitude towards IS/ICT adoption and use	DeLone, 1988 Caldeira and Ward, 2002 Thong et al., 1996
	b. Computer planning	DeLone, 1988
	c. Sophistication of control	DeLone, 1988
	d. User participation	Montazemi, 1988
	e. Intensity of requirements analysis	Montazemi, 1988
3. System characteristics	a. Type of computer used	DeLone, 1988
	b. Number of administrative applications	Raymond, 1985
	c. Interactive/online applications	Montazemi, 1988 Raymond, 1985
4. Internal expertise	a. CEO knowledge of computer	DeLone, 1988 Palvia et al., 1994
	b. Internal IS/ICT competences	Caldeira and Ward, 2002
	c. User computer literacy	Montazemi, 1988 Igbaria et al., 1998
	d. Presence of systems analysts	Montazemi, 1988
5. External expertise	a. Vendor's support	Yap et al., 1992 Caldeira and Ward, 2002 Thong et al., 1996 Igbaria et al., 1998
	b. Consultant effectiveness	Yap et al., 1992 Soh et al., 1992 Thong et al., 1996 Gable, 1989, 1991 Kole, 1983 Igbaria et al., 1998

Managerial sophistication relates to the mechanisms employed to plan, control and evaluate present and future applications (e.g. written documents, formalism of process, position of responsible individual and level of alignment with organisational objectives).

For the particular interest of our work, the "technological sophistication concept" developed in IS research was likely to help in understanding the conceptual framework of CAT tool adoption by providing the theoretical foundations used in this area to understand ICT adoption in the context of freelance translation businesses. In addition, Raymond and Paré developed and used an instrument based on the sophistication concept that can contribute to the development of the instrument measuring CAT tool adoption and the adoption of other ICT in the freelance translation business. In the absence of instruments to measure ICT adoption in the translation field, Raymond and Paré's instrument represented a useful and validated contribution to the measurement of ICT adoption to be tested in the freelance translation business context.

Part Two

Multilingual information and perspectives on ICT

"If we knew what it was we were doing, it would not be called research, would it?"

(Albert Einstein)

A research framework for Multilingual Information Management

In this chapter, a research model that addresses the issues behind the management of information in the multilingual context of freelance translators is presented within a wider framework of ICT adoption. This theoretical work is the outcome of an interdisciplinary review of research from the informant domains presented in section 6.1 and the work initiated by previous empirical research about freelance translators to provide a model of CAT tool adoption for this particular context (Fulford and Granell-Zafra, 2004; Fulford and Granell-Zafra, 2005; Granell-Zafra, 2006).

6.1 Informant domains

Today's advances in the technologies available to multilingual information professionals in general and to translators in particular aim to improve some aspects of their work, such as increasing their productivity or the quality of the translations they work with. The breadth of existing ICT and CAT tools, in addition to an increasingly wider availability of open source software for virtually all purposes, has provided more affordable and accessible tools to freelance translators today. However, there is not much evidence about the extent to which translation-specialist software, such as CAT tools, is actually being used among the freelance translator community. The literature reviewed in Part One of the book reported that CAT tool usage has not been studied thoroughly until recently, and that there is little evidence of its widespread adoption among translators.

On the contrary, the literature available in other information-related domains shows that ICT adoption has been studied in other areas and that IL has proved to be a suitable approach to facilitate MIPs' continuous professional development, thus helping to develop an interdisciplinary framework that addresses the issues in our work. The literature about ICT adoption in the context of SMEs in other sectors, such as manufacturing and accounting, is richer and provides an established and validated body of literature and instruments to draw on. Hence, this body of literature suggests that the following issues were worthy of investigation:

- models of ICT adoption in SMEs (Saarinen, 1989; Cragg and Zinatelli, 1995), which can help to identify the levels of ICT adoption in small translation businesses;

- factors that have motivated or inhibited the decision of ICT adoption in SMEs (Easton et al., 1982; Farhoomand and Hrycyk, 1985; Malone, 1985; Baker, 1987; Lefebvre and Lefebvre, 1988; King and McAulay, 1989; Cragg and King, 1993; Cragg and Zinatelli, 1995), which can help to identify the factors that motivate or inhibit the adoption of ICT by small translation businesses;
- success factors for ICT implementation in SMEs (DeLone, 1988; Montazemi, 1988; Palvia et al., 1994; Caldeira and Ward, 2002), which will help to identify the factors that lead to a successful implementation of ICT in small translation businesses;
- sophistication in the usage of ICT in SMEs (Benbasat et al., 1980; Cheney and Dickson, 1982; Ein-Dor and Segev, 1982; Saunders and Keller, 1983; Gremillion, 1984; Lehman, 1985; Mahmood and Becker, 1985; Raymond, 1988; Raymond and Paré, 1992), which will help to understand the evolution in the adoption and use of ICT in small translation businesses;
- the impacts of introducing specialist ICT (Lefebvre, 1996; Saarinen, 1996; Palvia and Palvia, 1999), such as adopting CAT tools into a freelance translator's working environment.

Therefore, the following domains have been identified as key informant domains for this research: first, information and communications technology and information systems; second, language and translation; third, information literacy; and fourth, small business management. Within these informant domains, the areas studied in Part One of the book include:

- Information and Communications Technology and Information Systems (Chapters 2 and 5)
  - measures for determining the success of ICT adoption and implementation in organisations;
  - stage models of ICT adoption in organisations;
  - systems integration and networking issues, including collaborative and workflow management issues;
  - Internet adoption and diffusion in organisations, especially in the small business sector.
- Language and Translation (Chapter 3)
  - software types employed by translators;
  - categorisation of translators (e.g. freelance, in-house);
  - translators' working practices and working environments;
  - translation workflow models.
- Information Literacy (Chapter 4)
  - ICT skills acquisition;
  - information use;
  - digital literacy;
  - lifelong learning.
- Small Business Management (Chapter 5)
  - ICT adoption strategies in small businesses;
  - ICT adoption factors in small businesses, including motivators and inhibitors;
  - the influence and role of the CEO in ICT adoption decisions;
  - small business performance measurements, including benefits realised from ICT adoption, together with problems encountered (adoption outcomes);
  - small business planning and business strategy formulation.

An overview of these informant domains and sub-areas identified is presented in Figure 6.1 below.

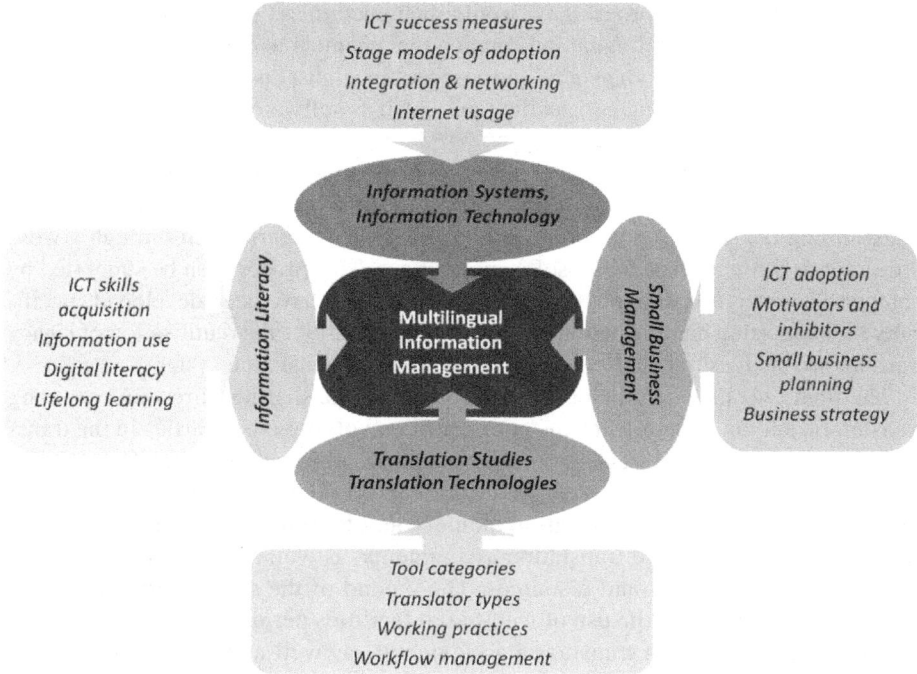

Figure 6.1 Informant domains

6.2 Development of the conceptual framework

Having identified the key informant domains, a conceptual framework needed to be devised to support the empirical exploration and analysis of the various factors, both internal and external, that might motivate or inhibit the adoption of ICT by small translation businesses, as well as to gain insights into training and the use of information, to enable the identification of successful adoption strategies, and the measurement of post-adoption firm performance.

First, the nature of the translation work was assessed, and the core translation processes identified and categorised. The tools used to support each of these processes were considered, and a progression of processes and tools established based on an increasing level of ICT sophistication (as reported in studies such as Nolan, 1973; Raymond and Paré, 1992; Cragg and Zinatelli, 1995; Joscelyne, 2003).

This progression begins with the process of text and document production and editing, in which general-purpose software tools, such as word processing software, text and web editors, and desktop publishing tools might be employed to produce, format and present translation assignments. In second place, typical business management processes are present to support the day-to-day operations of a small translation business, including communications with clients and providers (i.e. multilingual vendors, translation companies or agencies), office management duties, and budgeting and invoicing.

The next stage of the progression would then encompass the task of terminology management, involving bilingual information management to create terminology collections, with the use of both general-purpose and specialist databases being important here, as well as more language-specific tools, such as software for creating electronic glossaries, dictionaries or corpora.

Continuing the progression of increasing sophistication, the next stage can be termed the "translation creation level," i.e. the actual task of formulating the translation, and transforming the source text into target text, whether it is performed through a word processor or through a specialist-software interface. This process can be supported by computer-assisted translation software applications that have been developed specifically for supporting the translation process and that facilitate the reutilisation of legacy translations, such as translation memory and machine translation systems.

The final two stages in the progression focus on the sharing of resources among translators, and on wider electronic cooperation involving other parties in the translation process, such as clients, project managers, subject experts, colleagues and so on. The first of these stages can be deemed the "collaborative level" and entails sharing resources such as translation memory and terminological databases to facilitate teamwork on large translation assignments, as well as online collaboration through web-based tools and resources. The second of the stages, termed the "integrated level," involves the use of web-based facilities permitting the formation of global teams working on translation assignments, as well as the creation of electronic marketplaces in which translations are bought, translation services offered, and translation assignments received and transmitted. Included in this final level would also be an emphasis on a greater degree of integration of different types of software, such as the integration of terminology management software, CAT tools, MT and project management software, leading to a fully "interconnected" electronic environment that works as an information system in the context of multilingual information management.

This idea of an integrated environment, initially put forward by previous research undertaken by the author and Heather Fulford already pointed towards an integrated workstation that addressed ICT adoption in the context of freelance translators (Fulford and Granell-Zafra, 2004). This view is also shared by other scholars, such as Taravella and Villeneuve, who also advocate an integrated perspective of the information system behind the translators' activity and their information management function in what they call a "Language Information System" that helps translators to "share and disseminate information, in a way that usefully informs translation production processes" (Taravella and Villeneuve, 2013, p. 63). However, similarly to the approaches followed by research in Translation Studies (cf. Chapter 3), the "Language Information System" put forward by these authors is mainly focused on the core translation activity of these professionals and the interaction of translators with automated tools: "we will stick to [the term] automated interactive translation [...] [i.e.] translation executed with some use of machine translation, but considered as the result of collaboration between machine and human translators" (idem, p. 66) so that they "will routinely integrate machine translation tools, computer-aided translation tools, and passive language technology

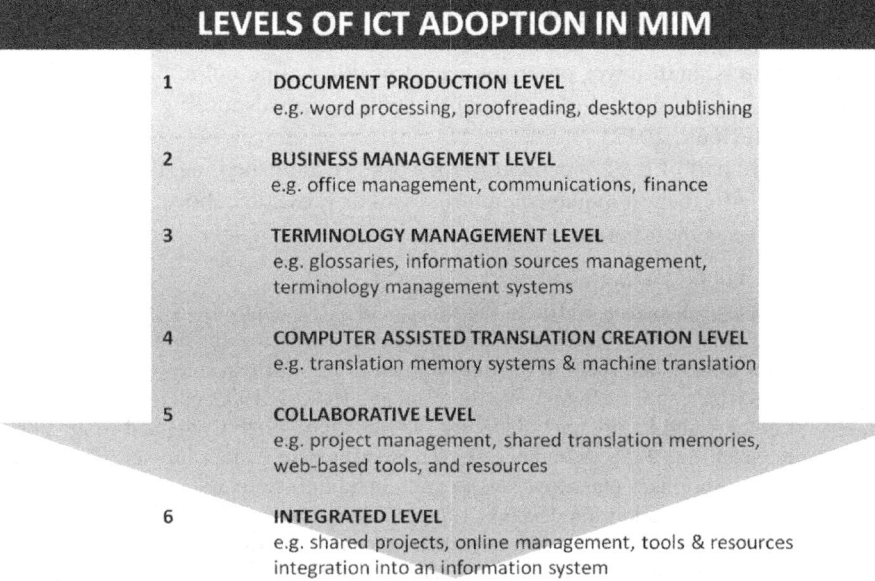

Figure 6.2 Overview of levels of ICT adoption in Multilingual Information Management

(e.g. online databases and shared glossaries), with their own human knowledge, in order to produce a translation" (idem, p. 66).

The workstation view of Fulford and Granell proposed in 2004 was broader in scope and encompassed translation-specific ICT, i.e. CAT tools and information resources, and general-purpose ICT that supported other functions of the multilingual information setting of freelance translators. Here, this proposal is reintroduced to provide an updated perspective of this environment fuelled by an IS and IL approach to understanding MIPs' needs for information and technology throughout all their professional activities.

An overview of this progression of processes and tool sophistication is shown in Figure 6.2 and further described in the next section of the chapter.

In proposing these six levels of ICT sophistication within the translator's working setting, there are three aspects that further expand the translator's workstation as defined by earlier proposals, such as that put forward by Melby (1992), as well as that outlined by Somers (2003b). First, the business management level, to acknowledge the freelance environment in which many translators work today, and thus to incorporate the range of tools they are likely to require to run their freelance operations effectively (office automation software). Second, to recognise and accommodate, through the collaborative level, the increasing emphasis on ICT developments on interconnectivity and networking, making it possible for freelance translators to work together on group translation assignments, to communicate electronically, to take advantage of social networks and Web 2.0 services, and to share the necessary terminology and memory resources to achieve this. Third, the integrated level, to acknowledge the growing trend of adopting more formal project management approaches to the undertaking of translation assignments in globally distributed teams, the tendency towards a cloud-computing-based framework that enables web-based applications, and newer practices based on the online collaboration of multiple (and usually large) networks of people, such as crowdsourcing (Howe, 2006; Kelly, 2009; Mesipuu, 2012).

Once the core part of the conceptual framework is established, i.e. the progression of multilingual information management processes as presented above, four additional components were incorporated into the framework.

- **Factors affecting ICT adoption**
  In order to assist translation SMEs in the process of ICT adoption, an understanding is needed of the factors that influence their adoption decisions. To this end, part of the literature review undertaken focused on an examination of prior research on ICT adoption decisions in SMEs (see for example Farhoomand and Hrycyk, 1985; Malone, 1985; Baker, 1987; Lefebvre and Lefebvre, 1988; King and McAulay, 1989; Cragg and King, 1993; Cragg and Zinatelli, 1995). A number of factors influencing adoption were identified, and then, in a preliminary classification, were grouped into motivators (factors that have a positive effect on an adoption decision) and inhibitors (factors that have a negative effect on an ICT adoption decision). Motivators might include CEO involvement, promises of increased productivity and competitive pressure. Inhibitors might include such factors as lack of IS/ICT knowledge, lack of technical support and rapid technological changes. Some of these factors are also related to the training received and how information literacy is developed.
- **Business strategy and ICT strategies**
  The purpose of this component of the conceptual framework was to identify the business and ICT strategies that MIPs are putting in place, and to consider the effects these strategies might have on a their ICT adoption decisions. From an IL angle, strategy should also take into consideration continuous training and how professional development can be achieved.
- **Firm performance**
  Having considered those factors that might influence ICT adoption decisions, the effects that ICT adoption has on the performance of a MIP should also be examined, including the benefits that might be derived from ICT adoption and the problems that might be encountered following adoption. Again, part of the literature review undertaken permitted the identification of a number of firm performance measures that can usefully be drawn upon and applied to the translation sector, specifically to measuring and analysing the various outcomes of translation tool adoption (Caldeira and Ward, 2002; Cragg and King, 1993; DeLone, 1988; Cragg and Zinatelli, 1995; Montazemi, 1988; Raymond, 1989; Thong et al., 1996). These measures include productivity improvements, increased revenue and time saving.
- **Information literacy**
  Finally, as stated in Chapter 4, an IL perspective provides a useful aid to understanding information-related needs, being aware of how to tackle them, and staying up to date through a lifelong learning approach. Thus, IL embraces all the activities involving the use of tools and resources to manage information and analysing the needs to be met to face the ever-changing landscape of information and technology.

A diagrammatic overview of the conceptual framework outlined above is presented in Figure 6.3.

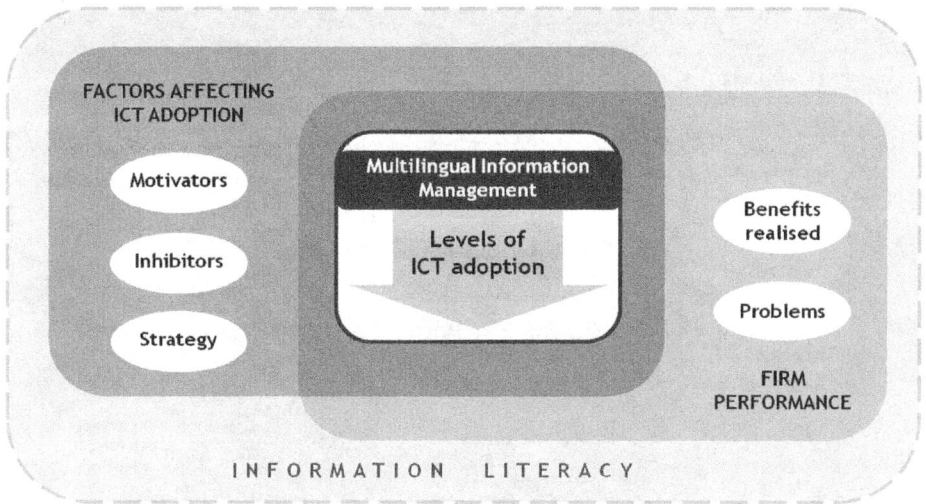

Figure 6.3 Overview of the conceptual framework

6.3 A research model of Multilingual Information Management

Based on this theoretical framework, a research model was developed to analyse the ICT setting of MIPs and, in particular, the adoption of CAT tools by freelance translators as the most specialised type of translation-dedicated software, to determine the relationships of such specialist software with the factors that may determine the adoption decision, and to examine the impacts that all this may have on the translators' workflow. In the discussion about the tools and resources used by freelance translators and the translator's workstation in Chapter 3 and in previous works (Fulford and Granell-Zafra, 2004; Fulford and Granell-Zafra, 2005) a number of activities involving MIM and ICT have been identified. In this part of the book, this activity view is detailed to provide the basis of the research model.

This model comprises the determinants that are likely to affect the adoption of ICT tools, with a focus on CAT tools, the support provided by ICT within the Multilingual Information Management context of translators, and the impacts that specialised ICT may have. The determinants that may affect the decision to adopt CAT tools are represented by the "Translator characteristics," the "Translation business characteristics," and the translators' "Perceptions of specialised ICT" (i.e. CAT tools). The core part of the model is presented in the form of ICT-dependent activities that are part of translators' workflow (i.e. document production, information search and retrieval, business management, translation creation, communications, and marketing and work procurement). ICT success is represented by the impacts that the adoption of specialised ICT may entail, such as the benefits realised or problems that CAT tools may cause to the translators' workflow.

Figure 6.4 displays the model and each of the components is described in detail in the following sections of the chapter.

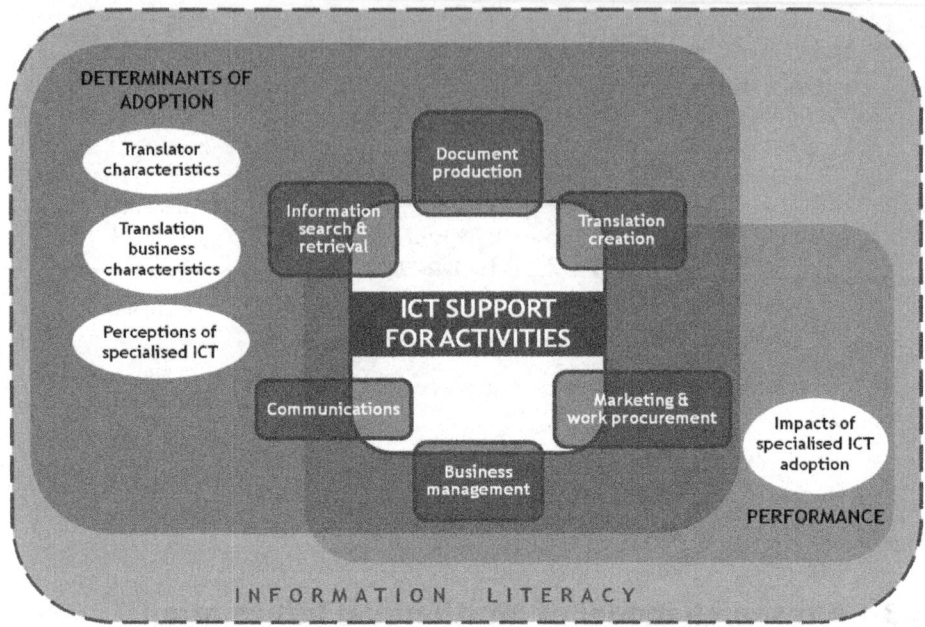

Figure 6.4 Research model: ICT adoption by freelance translators

6.3.1 ICT support for activities

Section 3.3 in Chapter 3 showed how the literature about ICT and translators has traditionally focused on discussing the tasks and sub-processes surrounding the core translation activity, usually categorising technologies according to their level of automation of tasks (e.g. Hutchins and Somers, 1992). Austermühl's process-oriented approach (2001) proposed three levels at which translation tools may help translators (at the translation workflow management level, at the linguistic and cultural transfer level and at the automation level) which expanded the traditional understanding of translation technologies ranging from human-driven tools to fully automatic tools. However, as noted by other authors there are a number of activities supported by ICT that freelance translators must undertake [for example, Locke (2005) cites budgeting, pricing, and hardware and software acquisition]. Fulford and Granell (2005) further extended Locke's list of activities making up the freelance translator's workflow to include marketing, work procurement, communication/client liaison, bookkeeping/ financial management and billing/invoicing (also highlighted by other authors such as Varona, 2002).

This "activity view" of the freelance translator's workflow provided the basis for researching CAT tool adoption within the ICT and business context of freelance translators and is revisited here. In this work, the emphasis is on investigating, not so much the degree of automation that translators are willing to permit into their core translation activities, but rather on the range and types of information and communication technologies they are adopting to support them in each of the individual activities that form a constituent part of their overall workflow, as well as the extent and intensity to which ICT are used. This broader view of activities draws on Austermühl's "process-orientated" view of the translator's workstation. He suggests that in this process-orientated view, the ICT that translators use must encompass both the notion of "translation as a business" as well as translation "as a linguistic and cultural process" (Austermühl, 2001, p.11). This process-orientated approach is, however, not specific to the freelance working environment, and so does not include important components of the freelancer's role as a form of business, such as marketing and financial management. The activity view adopted here extends Austermühl's approach to incorporate these components, thereby providing a more comprehensive view of the activities making up a freelancer's workflow and contributing towards the development of a Multilingual Information Management System (MIMS), defined in more detail in Chapter 9.

Drawing on the model presented in Fulford and Granell, (2005), an updated and detailed description of freelancer activities is provided in the following sections. Figure 6.5 shows an overview of the activities and ICT model and it is followed by a discussion of each of the activities and a table summarising the tasks involved in each, the type of technology supporting it, and some examples of applications and resources.

Figure 6.5 Multilingual Information Management activities

6.3.1.1 Document production

Being the modern version of typewriters, it is fully understandable that word processing software has become the main tool for composing, editing and adding format to text-based documents. This has also been the case for translators, whose main working interface has traditionally been a word processor since the age of mechanical and electronic typewriters, usually Microsoft Word or WordPerfect. Today, apart from these and other commercial options, open source applications such as OpenOffice Writer and LibreOffice Writer are becoming more and more popular among personal and professional users as differences in the features and supported file formats have practically disappeared between the former and the latter (Díaz Fouces and García González, 2008, p. 3). In addition, many individuals and organisations are increasingly using the Internet for storing, presenting and disseminating documents, and web-based word processing applications, such as Google DriveDocs or Zoho Writer, are also becoming more popular due to their advantageous online features, such as free online storage, access from most web browsers and mobile devices, ease for sharing files and collaborating among several users and integration with other online tools. Typical features of word processing software include text editing and automating actions (such as generating text page numbers, dates, automatic text correction, tables of contents etc.), spelling and grammar checking, looking up built-in thesaurus and dictionaries, using templates and styles to automate layout, adding tables, graphs, images and other graphical objects, and exporting and saving contents to several file formats (such as HTML or PDF).

Another interesting feature of word processors is their capability to define and execute macros (short for for "macroinstruction") that reduce repetitive tasks by specifying a sequence of actions, saving the procedure, and running it each time it is required by a keystroke combination or a customised button. For example, Microsoft Word includes built-in macro development through Visual Basic for Applications (VBA), an implementation of the Visual Basic programming language, to create sets of functions, automating processes, or interacting with other applications through the Word interface (such as translation memories or reference managers).

Besides the increased capabilities of word processors to create documents, other types of software are specially addressed to deliver professional layouts, work with typographies, images and graphics, such as Desktop Publishing (DTP) applications or slide show presentations software. Examples of these are provided in Table 6.1.

Likewise, much corporate marketing and promotional material is mainly delivered online today, and information providers, such as government agencies and other such bodies, rely on the Web for document and information dissemination. A translator today would arguably be wise, therefore, to gain some understanding of, and become familiar with, web publishing software, as well as the more usual word processing and DTP packages. Such software might include WYSIWYG (What You See Is What You Get) HTML editors such as Dreamweaver or KompoZer.

6.3.1.2 Information search and retrieval

As multilingual information managers, translators spend a good deal of their time looking for information. The purpose can widely vary. To name a few examples, it

Table 6.1 Document production activity

Activity	Tasks	ICT Support	Examples of applications
Document production	Creating target text Adding format to target text Overtyping source text with target text Adding/editing metadata Editing graphical assets	Word processing software	*MS Word, WordPerfect, OpenOffice Writer, LibreOffice Writer, Google Drive Docs, Zoho Writer*
		Graphical/presentation software	*MS PowerPoint, Apple Keynote, LibreOffice Impress, Google Drive Slides, Zoho Show, Prezi*
		Web publishing software	*Dreamweaver, KompoZer*
		Desktop publishing software	*QuarkXpress, InDesign, PageMaker, Scribus*

may help them to find the accurate term in a particular language for a specialised medical technique, to make a decision about the most suitable style and jargon to be used in a TV show about gardening and DIY, or to build a glossary or a terminological database with the appropriate translations and phrases for a PlayStation video game, according to their client's requirements. Hence, this activity involves a number of tasks, ranging from locating background and reference material to help with the acquisition and verification of subject field knowledge, to the identification of specialist terminology in various languages. Traditionally, translators have relied heavily on paper-based terminological and reference resources, and on the help provided by fellow translators or subject field experts (Fraser, 2001). Reliance today is also considerably placed on online information sources. The Web has considerably increased translators' choices in terms of the breadth of sources they can look up and of the ways they can consult them, whether material resources or physical people.

Included among these are websites, such as company or institutional websites, or sites offering information about particular subject fields, blogs fed by subject field experts, online reference works, document archives, digital libraries, online corpora, specialist databases, online dictionaries, glossaries and terminology databanks.

Given the vast amount of online resources and of information overload provided by the Internet and the digital age (Aikat and Remund, 2012, p. 28), the way of accessing information is also key to multilingual information users; Internet search engines are the most direct and simple way of seeking information, although not necessarily the most valid in terms of accuracy or relevance. For this reason, multilingual information users should also be able to access information through more specialised channels, such as via specialised databases, making use of intelligent search tools, using information filtering services, and utilising terminology management software.

With regard to the information obtained from subject field experts or fellow translators, far beyond the channel opened by email a couple of decades ago, ICT today provides a number of online services and tools to contact individuals and follow the information they share on the net. These online communications include, for example, blogs, online forums, discussion lists, online communities, content curation customised gateways and social networks.

Examples of the resources and ICT discussed in this activity are presented in Table 6.2.

In order to exploit these resources and ICT effectively, translators need to become proficient in addressing their information needs by selecting the most suitable search strategy to support complex information retrieval tasks. As stated in Chapter 4, this strategy is a process that includes (1) being able to identify the information needed, (2) being aware of the information sources available, (3) selecting the optimal approach to find the required information, (4) being capable of accessing the information sources and retrieving data, and (5) being able to evaluate the results of the searches and selecting the most appropriate information, or identifying invalid results and looking for alternative approaches. Inefficient online searching by translators will be costly for their businesses, both in terms of time and money. Given the importance of online resources to translators, investment in acquiring the knowledge and skills needed to become adept at using online search facilities will be very valuable in the early stages of their training and professional development, as well as during their lifelong professional development.

6.3.1.3 Translation creation

The central activity of a translator is, obviously, undertaking the language transfer between the two languages involved. Very briefly, this involves the formulation of a translation, its revision, editing and proofreading. Although translators have traditionally relied on manual methods and basic IT support to produce their translations, fundamentally using a word processor and other general-purpose office software, the technological developments over the last two decades have provided a growing array of ICT to support the translation task and automate part of the process, some of which have specially been conceived to this end (cf. Chapter 3).

Among the range of specialised-purpose tools and technologies involved in the process of creating translations, four groups of support can be distinguished, namely, CAT tools, proofreading and quality assurance tools, localisation tools, and machine translation technology.

In the CAT tools group, a number of sub-processes are supported by ICT, translation memory and terminology management software being the most important functionalities. The former enables an effective reutilisation of previously translated text at term, phrase or segment level, usually referred to as leveraging; fully or partially, often called "100% matches" and "fuzzy matches"; and from personal prior translations or from legacy translation memories provided by translation companies. The latter helps to ensure terminology coherence and consistency throughout a document, within a set of documents from a project, or between old versions and newly revised texts. Each piece of software provides its particular technology for automating searches, feeding TMs and terminology databases, interacting with external resources, or customising its interface to the user's convenience. A more detailed definition is presented in section 3.1.

Table 6.2 **Information search and retrieval activity**

Activity	Tasks	ICT Support	Examples of applications
Information search & retrieval	Locating background and reference materials	Internet search engine	*Google, Yahoo, Bing, Creative Commons Search, Google Scholar, Scirus*
	Using databases	Electronic encyclopaedia/ reference work	*Encyclopaedia Britannica, Wikipedia, MedlinePlus Medical Encyclopedia, Gale Virtual Reference Library*
	Locating client company information		
	Identifying terminology	Electronic dictionary and glossary	*Diccionario de la Real Academia Española (DRAE), Wordreference, Merriam-Webster, yourDictionary.com, OneLook, ProZ.com term search, AcronymFinder*
	Locating definitions of terms		
	Finding examples of terminology usage	Terminology databank	*Eurodicautom/IATE, CILF, Microsoft Language Portal*
		Text corpora / document archive / online translation memories	*CREA, British National Corpus, New Scientist Archive, ACM Digital Library, Redalyc, Linguee, Glosbe, European Commission multilingual translation memories, TAUS Search*
	Finding parallel texts		
	Using multilingual corpora		
	Managing personal terminology collections	Terminology management software	*SDL MultiTerm, Star Transit TermStar, Déjà Vu X TermBase, Lexicool Lingo*
		Online specialised databases	*Medline/PubMed, IMDB, Factiva, ISI Web of Knowledge, IEEE Xplore, Scopus*
		Electronic library	*The British Library, Biblioteca Nacional de España, National Library of Medicine, SciELO*
		Database software	*Microsoft Access, OpenOffice Base, FileMaker*
		Intelligent search engines and applications	*Copernic Agent, IntelliWebSearch, Funduc Search and Replace, InfoRapid Search and Replace*
		Content curation/ filtering tools	*ScoopIt!, Paper.li, The Tweeted Times, BBC World Service, iTunes, Google Reader†, Feedly, Flipboard, Pulse*
		Podcasting	
		RSS	

In addition, CAT tools, usually available as bundled software or suites, also include a growing number of features to support the translation task, either from an integrated interface, as in SDL Trados, memoQ and most of current CAT tools, or as a package of applications that interact with MS Word or other programs, such as MetaTexis or older versions of Trados and Wordfast. Some of these features are, for example, extracting terms, aligning texts to produce translation memories from bilingual texts or previous translations, editing texts in tagged formats, such as HTML, XML, or XLIFF, working with different document formats through import/export functions, sharing TMs through servers, performing global changes on documents and TMs, and providing a number of analysis functions to pre-process files against TMs and obtaining an estimation of partial and full matches, calculating statistics, and so on.

Proofreading features are also present in most CAT tools, although word processors have their own built-in language revision tools for spelling and grammar checks. There are also dedicated language tools and web-based applications, such as Stilus, for this purpose. The same happens with quality assurance tools. Almost all CAT tool packages have some sort of function to perform QA checks in single or multiple files, such as spelling checking, untranslated text, mismatches between source and target text, or tags checking. However, there are also dedicated tools to this end, such as ApSIC Xbench.

The group of ICT focusing on localisation serves most of the tasks that CAT tools support, although localisation tools are specifically designed to deal with all the technical and format difficulties, among others, present in the process of translating digital contents and products (mainly software, websites and video games) and adapting them to the language, culture and non-textual conventions of a particular local market (usually called a locale) in terms of geographical area, language and culture (Dunne, 2006, p. 4). A more detailed discussion of the concept of localisation can be found in the volume edited by Dunne, in particular in the introduction and in the chapter authored by Folaron (2006), (as well as in the previous works of Esselink, 2000; O'Hagan and Ashworth, 2002; Pym, 2004).

A few examples of applications are presented in Table 6.3, including localisation suites such as Catalyst or Passolo; localisation tools for specific purposes, like Cats-Cradle for localising websites and Image Localization Manager for localising image files; or tools for particular file formats, such as Poedit (.po files) or Logoport XLIFF Editor (XLIFF files) (Mata Pastor, 2008).

Machine translation technology today is present in various ways. It is available as an online service, e.g. Google Translate, or as a software package, e.g. Systran, to those users who do not seek for a necessarily high quality translation of their texts, such as individual users and businesses looking for the quick gist of a text or for a rough translation of a potentially interesting text. In addition, machine translation, once considered to be the future solution for multilingual communication needs (cf. Chapter 3), has rather served to the purposes of human-mediated professional translation since the end of the 1990s, either by enhancing the technology behind some translation memory systems (Melby, 2007; de la Fuente, 2012), or by producing translations to be post-edited and revised by humans, a growing trend in the 21st century (TAUS, 2009; Garcia, 2011; Melby et al., 2012), particularly in those fields where language can be

Table 6.3 **Translation creation activity**

Activity	Tasks	ICT Support	Examples of applications
Translation creation	Formulating translation Re-using legacy translations Revising and adapting TM fuzzy matches Ensuring text coherence at term and phrase level Aligning text from previous translations Working with tagged text formats Sharing TMs with other translators Checking spelling & grammar Performing quality assurance checks Using format filters Post-editing MT output	CAT tools Translation Memory Terminology Management Proofreading tools QA tools Localisation tools Machine translation	*SDL Trados, Déjà Vu, Star Transit, Wordfast, Wordfast Anywhere, memoQ, OmegaT, MetaTexis Word processing built-in language tools, ApSIC XBench, Stilus Alchemy Catalyst, Passolo, CatsCradle, Image Localization Manager, Google Translator Toolkit, Poedit, Logoport Tools, Logoport XLIFF Editor, Microsoft Helium Reverso Pro, Systran, Internostrum, SALT, Google Translate, Babylon*

controlled (Aikawa et al., 2007; Fiederer and O'Brien, 2009) or in certain settings where human translation is not possible due to economic (Bowker, 2010) or high volume reasons. Similarly to the use made of MT by TM (Garcia, 2005; Lagoudaki, 2008) for providing a draft MT output to assist in the translation of text not found in the translation memory, there is even research that reports on the usefulness of MT for freelance translators, at least on an occasional basis, for identifying translation suggestions (Fulford, 2002a), for looking up short sentences (Champollion, 2003), or for starting to work on their final translation from a rough draft (O'Hagan and Ashworth, 2002, p. 43).

6.3.1.4 Communication

Being involved in multilingual information management in today's digital world means that you need to be "connected to" and in permanent communication with a number of people and through a wide variety of channels.

For instance, freelance translators must liaise with their direct clients, contact subject experts, negotiate job requests with translation agencies and manage queries with the project managers of large projects they are working on. In addition, communicating with colleagues is essential for translators, since "fellow translators can help provide a rich source of assistance, whether this be in the form of work provision or job sharing, assistance with terminological queries, provision of translation revision services, proofreading services, or more informal (but nonetheless highly valuable) moral support and encouragement in an otherwise somewhat isolated profession" (Fulford and Granell-Zafra, 2008, p. 11). As freelancers are running their own business, so they may also be in need of contacting business advisers such as accountants, legal advisers and ICT specialists.

Given the facilities provided by today's digital environment and the global market in which translators work, geographical distances are no barriers to them. ICT supporting communication tasks entails using a number of channels, from direct one-to-one or one-to-many email and instant messaging facilities to more public and social communication networks, such as online discussion groups and forums, communities around very particular topics, or publicly available services for exchanging and disseminating information online. Examples of these electronic communication facilities and services are provided in Table 6.5 and discussed below.

The successful use of such ICT depends on translators being fully acquainted with the various channels available to them for transferring and exchanging files electronically, such as exchanging source and target texts with clients, sharing translation memories and glossaries with other colleagues, or sending translation deliverables to project managers. This requires knowledge of email clients; Instant Messaging (IM) applications; and file transfer and sharing tools and services, such as FTP (File Transfer Protocol) clients, other technologies for exchanging files, like P2P (peer-to-peer) clients, and cloud computing file hosting services (such as Google Drive or Dropbox) that enable storing and sharing files on the web.

Although email is not a sophisticated type of ICT and everyone is familiar with it, an efficient management of email accounts and messages is deemed essential within any form of business (Sainz-Aloy and Soy-Aumatell, 2011), especially for those professionals working with information. Some examples for optimising email management are organising messages in folders or tags according to their relevance and the actions required, keeping the inbox as empty as possible, minimising unwanted email through spam filters, deleting unnecessary messages, making use of email clients' functionalities to automate message classification or filtering messages and performing searches effectively. Similarly, asynchronous electronic communications in this business environment should take special care about the way language is used and follow suggestions on "netiquette," such as those provided by Shea and Shea (1994), Scheuermann and Taylor (1997) or Sainz-Aloy and Soy-Aumatell (2011). They include hints about how to formulate electronic messages correctly, such as always replying to messages, writing clear, concise and informative subject lines about only one topic, using polite and respectful language and spell-checkers to avoid typos or mistakes originated by quick replies, communicating clearly and concisely, writing well-structured messages, minimising long conversations by email, double checking recipients and limiting the use of "Reply all."

Many multilingual translation projects are carried out collaboratively, in particular those involving more than two languages or comprising a large volume of words. Thus, as mentioned earlier, translators need to collaborate with their clients, with the LSPs coordinating their projects, with cross-discipline teams and with fellow translators. For simple collaboration tasks among a small number of collaborators, email attachments or IM file transfers can fulfil the job. However, alternative solutions are required due to the limits of the size of files that can be transferred and the inconvenience or impossibility of physically carrying the data in removable data storage media, such as USB drives, CDs or DVDs. Large files and sets of files are usually transferred using file transfer tools and web-based file hosting services. An FTP client, for example, is used to connect to the FTP server hosting the files and transfer them, and even more easily, most web browsers can access FTP servers by entering the URL in the location bar with "http" replaced by "ftp." On the other hand, cloud storage services, such as Dropbox, Google Drive or SkyDrive, are gaining more and more popularity (Drago et al., 2012) since they are relatively easy to use, can be accessed from any device, and are rather inexpensive or free to use to a certain extent.

Another important issue is how the social Web has changed the way people communicate. Now, a large number of online tools and platforms allow people to interact and to share content and are fed by user-created content, such as participative encyclopaedias (e.g. Wikipedia), social bookmarking platforms (e.g. Delicious, Connotea, Diigo), photo and video sharing platforms (e.g. Flickr, YouTube, Pinterest), text or visual documents platforms (e.g. Slideshare, Prezi, Issuu, Wikispaces), or microblogging platforms (e.g. Twitter). Such tools enable users to share content (photos, text posts, documents, videos, bookmarks, etc.); to express their opinions through comments to others' contributions; to connect with other users with common interests; or to add free-text tags or keywords to content. Multilingual information professionals are no strangers to Web 2.0 and it has become commonplace to them, opening a whole new environment to online communities of translators with common interests where they can exchange ideas, discuss professional issues, share job requests, find collaborators for their projects, make terminology queries or just network with fellow translators. In addition to gathering around specialised networks (e.g. Traditori, Langmates, Proz, TranslatorsCafe.com), they have also expanded their online social activity to other general-purpose services of the 2.0 scenario, such as special interest Facebook or LinkedIn groups, or even ad-hoc topic-focused discussions generated within social networking services like Twitter, just through hashtags (a hash symbol "#" followed by a keyword) (Bruns and Burgess, 2011).

Among the Web 2.0 services related to information management there is a growing trend towards the so-called "digital content curation" (Borgman, 2007), briefly consisting of the process of selecting, collecting, organising and displaying digital information relevant to a particular topic (Nielsen and Hjørland, 2014) with the aim of preserving and adding value to this body of digital information and knowledge for current and future use (Joint Information Systems Committee 2003). This process of filtering contents, reviewing and broadcasting them, also present to some extent in general social networks like Twitter (Popova, 2011), has led to the popularisation of web-based content curation tools, such as ScoopIt!, Flipboard and Paper.li, where users share and discuss their "discoveries." These content-customised gateways gather contents about virtually almost every interest, but when it comes to specialised topics, they can be a useful knowledge catalyst and a valuable way to access expert content, leading to the creation of "content curation communities" (Rotman et al., 2012, p. 1092). If such communities are focused solely on curating scientific data, then community data systems and related efforts can be considered "curated databases" (Buneman et al.,

2008). As Rotman et al. acknowledge, "scientific progress is, in large part, dependent on the development of high-quality shared information resources tailored to meet the needs of various scientific communities" (ibid). As a sample of the dimension of these tools for specialised topics, a search of the keyword "content curation" in ScoopIt! resulted in 342 related topics (i.e. content curation communities), each of which with a number of followers and including 34,733 "scoops" (i.e. posts published by the people managing each of the topics).[1] For example, as of 12 March 2014, one of the results was the community "Content Curation World" (*http://www.scoop.it/t/real-time-news-curation*), created in January 2011 and followed by 4974 people, where 1551 scoops had been published by then, accumulating over 592,000 visits, and that originated 64,507 reactions (i.e. commenting, acknowledging, or sharing the content published).[2] Some of the functions, other than publishing and commenting, are filtering posts by keyword, suggesting content to the topic curator, sharing posts and flagging inadequate content.

6.3.1.5 Marketing and work procurement

Any freelance information professional, including freelance translators, needs to pay special attention to the promotion of their services and the continuous search for new clients or contract opportunities. Although it might not be a central activity to their business, efforts are likely to be made to obtain translation work, be it in the form of new clients who contact them directly, responding to job requests offered by translation agencies, or placing bids for contracts in online translation markets.

In addition to traditional opportunities for promoting their business, such as paper-based directories, business cards and promotional brochures or leaflets, a freelance translator competing in the global market today needs to develop an online presence. This means, from being listed in relevant online directories and listings of translation services providers (see examples in Table 6.5), to building a strong online presence powered by social media. In between, there are a number of ways to use the web as a marketing channel, such as having a website (which can widely range from showing basic contact details, services provided and a résumé to a full business website that delivers a professional image, detailed information about the services and processes followed for projects, subject-field related news, a portfolio or samples of projects, and any additional related add-ons, such as a blog or links to profiles in social networks); creating an online profile in professional social networks (including background information, skills or projects involved, like those in LinkedIn); writing about interesting topics related to the profession on a blog, social networks or other social media (see Table 6.4 for some examples under the ICT support labelled "Social bookmarking and

---

[1] An indicator of the growing popularity of content curation is that this search, performed on 12/04/2014 (*http://www.scoop.it/search?q=content+curation&type=topic&page=1&limit=24*) produced more than double the results than on 22/08/2013: 143 related topics, including 23,069 posts.

[2] Another illustrative result with even more followers was the community "Social Media Content Curation" (*http://www.scoop.it/t/social-media-content-curation*), created in April 2011 and followed by 5442 people, where 1213 scoops had been published by then, accumulating over 130,300 visits, and originating 37,338 reactions.

Table 6.4 Communication activity

Activity	Tasks	ICT Support	Examples of applications
Communication	Liaising with clients Networking with colleagues Collaborating with other translators Managing queries with project managers Communicating with subject field experts Filtering and disseminating specialised information among communities of knowledge	Email Transferring and sharing files Electronic mailing lists Online discussion groups / forums Content curation tools and repositories Social networks and online communities of translators Instant Messaging (IM) Social bookmarking and online exchange and dissemination of information	Mozilla Thunderbird, Microsoft Outlook, Gmail, Apple Mail FTP clients (e.g. FileZilla, Cyberduck) P2P-based tools (e.g. Skype, BitTorrent) File hosting services (e.g. Google Drive, Dropbox, Box, SkyDrive) LANTRA-L, the LINGUIST list, Trag, IGDA Locsig TranslatorsCafe.com, Proz, Google groups, Yahoo groups, Facebook groups, LinkedIn groups ScoopIt!, Listly, Paper.li, Flipboard Open Language Archives Community (OLAC) Twitter, Facebook, LinkedIn, Xing, Plaxo, Traditori, Langmates, Proz, TranslatorsCafe.com Skype, ICQ, Line, WhatsApp, Google Hangouts, Windows Live Messenger, AIM, Pidgin, Jabber, Yahoo Messenger Diigo, Delicious, Scribd, Connotea, Slideshare, Prezi, Issuu, Wikispaces, Flickr, YouTube, Pinterest

online exchange and dissemination of information"); becoming an active member at group-specific discussions in online networks; and filtering and commenting online resources through topic-related gateways (see the discussion on "digital content curation" in the previous section and examples in Table 6.4). The high degree of connectivity of Web 2.0 services definitely affects marketing (Hanna et al., 2011), and this can act as a work procurement booster (Ardichvili et al., 2003) from the very moment that

Table 6.5 **Marketing and work procurement activity**

Activity	Tasks	ICT Support	Examples of applications
Marketing and work procurement	Promoting translation services Searching for clients Bidding for translation contracts Building an online presence Being connected through social media Disseminating knowledge	Personal web site/ blog (online presence) Online social networks Online marketplaces Content curation customised gateways Social bookmarking and online exchange and dissemination of information	*Web editors Blogging services (Wordpress, Blogger) Online profile services (about.me, LinkedIn) Twitter, LinkedIn, Xing, Traditori, Langmates Foreignword.biz, Proz. com, Aquarius, TranslatorsCafe.com ScoopIt!, Listly, Paper.li, Flipboard Diigo, Delicious, Scribd, Connotea, Slideshare, Prezi, Issuu, Wikispaces, Flickr, YouTube, Pinterest*

one becomes part of a vast and active professional network where someone may need to quickly find another colleague to collaborate on a project or to do a job request out of their scope.

Similarly, if online translation markets are to be used for procuring translation jobs, reputation matters (Yoganarasimhan, 2013). Except for those projects where a client posts a request and auction mechanisms are used to match buyers and sellers, clients will look beyond the most convenient price for the service they are looking for. Once they receive a list of potential freelancers, or a number of quotations, that can do the job they are asking for, they will view the proposals and eventually narrow their choice down according to the freelancers' rating, their experience, their rates and other assets, such as those provided by an online presence as stated above. There are general-purpose freelance marketplaces, such as Elance, Guru or Freelancer, but also translation-specific ones, such as Proz (www.proz.com), Foreignword (foreignword.biz), Aquarius (aquarius.net), and TranslatorsCafe (translatorscafe.com).

All these ways to be present online can seriously contribute to increase translators' visibility and to be connected to their professional world.

6.3.1.6 Business management

The management of a freelance translation business, just like any other small business managers and freelance information professionals, requires efficient and effective business administration. This involves a number of functions in common with most businesses, such as client and contact data management, quotation preparation and submission, invoicing, accounting and financial management, and management of IT equipment. In addition, there are some functions closely related to other multilingual management activities, such as data management and project planning and management. Table 6.6 presents ICT supporting each of these tasks and several examples of applications.

Data management tasks consist of dealing with file and document organisation, administration, and archiving and sharing that needs to be undertaken. Due to the high volume of information within a translation business, many of these tasks are closely

Table 6.6 **Business management activity**

Activity	Tasks	ICT Support	Examples of applications
Business management	Data management (organisation, administration and archiving/ sharing) Client and contact data management Project management Contract quotations Billing/invoicing Financial management	Document management systems Content Management Systems (CMS) Reference Management Software Data archiving and sharing System backup Project management Database software Spreadsheet software Accounting/ bookkeeping package Office productivity tools	*Windows SharePoint Services Alfresco, Drupal, Joomla!* *Document imaging tools (OCR / PDF tools)* *Zotero, EndNote, Refworks, Mendeley, Diigo, Evernote FTP clients (e.g. FileZilla, Cyberduck) File hosting services (e.g. Google Drive, Dropbox, Box, SkyDrive) Disk image and virtual drive applications (VMWare, VirtualBox, ISO images) Project Open, Globalsight, Translation Office 3000, Tom's Planner – Project Planner, WebBudget MS Access, FileMaker, OpenOffice Base MS Excel, OpenOffice Calc, Google Docs/Drive Sage, QuickBooks, Zoho Invoice, Factucont Security tools (antivirus, malware, spyware, firewalls) Mobile solutions (Portable Apps) Online calendars (Google Calendar, iCal, Kontact)*

connected to the first three activities above, i.e. document production, information search and retrieval, and translation creation. ICT supporting data management, such as document management systems, vary considerably, from integrated systems, such as Windows Sharepoint Services, to individually tailored solutions provided by traditional database and office software.

Apart from keeping records of all the files about clients, projects or suppliers, multilingual information businesses need to take special care of the language-specific resources they develop, such as terminology collections and translation memories. Regardless of the ICT used, be it a CAT tool or a glossary in a spreadsheet, information resources need to be organised so that they can quickly be identified, retrieved, reutilised and kept up to date.

Dealing with all the files and online resources that are used as information sources by translators is an important issue for an information-based business (see section 6.3.1.2 on information search and retrieval) that is also closely linked to how data are managed. Given the digital nature of most documents, some ICT conceived for other specific purposes, such as Content Management Systems (CMS) or Reference Management Software, can also be efficiently used to manage and keep records of digital information.

In relation to these issues it is also necessary to briefly discuss how the assets of a translation project are organised and managed. This is particularly important if we focus on localisation projects, where the number of files is considerably larger, translation teams usually involve more than one person, or where deadlines and planning is tightly linked to a number of deliverables. Some examples of available ICT for translation projects are Project Open, Globalsight or Translation Office 3000.

6.3.1.7 An extended view of MIM activities

The outcome of the extended core element of the model is presented in Figure 6.6.

6.3.2 Determinants of ICT adoption

In order to assist MIPs when adopting ICT, a greater understanding of those factors that affect the decision and, therefore, the process of adoption is needed. Hence, one of the focuses of the literature reviewed was the identification of previous research that had examined these factors. A number of studies have researched factors affecting the adoption of ICT in the information systems domain (c.f. Chapter 5, section 5.2.3.). After the initial analysis of the factors, a number of items were collected from the works of various authors, and initially classified into motivators (factors that affect positively ICT adoption) and inhibitors (factors that affect negatively ICT adoption). These items and their sources are exhibited in Tables 5.2 and 5.3, respectively.

As other similar SMEs that use ICT, MIPs are advised by previous research and professional associations to plan and define their requirements for ICT (Proudlock et al., 1999). However, this advice is generally derived from the ICT success in large firms, which have a different hardware, software and support environment from the micro business context of MIPs. This advice needs to be tested in the translation micro business context to help determine, for example, whether a written document

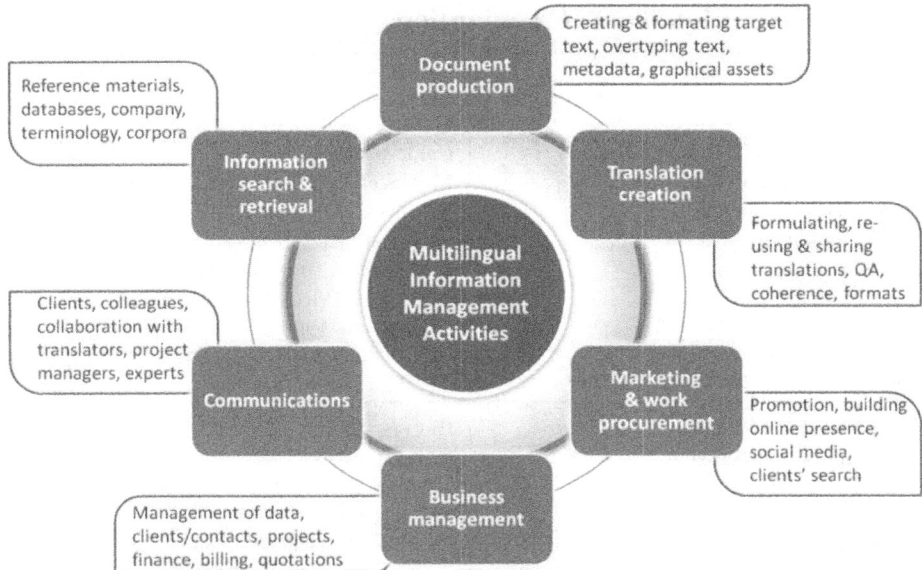

Figure 6.6 Multilingual information management activities

regarding ICT implementation is helpful; whether a particular level of ICT adoption is intended to be achieved directly, or a lower level of ICT should be adopted and then progressively move to a higher level of ICT adoption; whether the resources available will be enough to adopt a particular level of ICT successfully; and whether the degree of sophistication needed to adopt a particular level of ICT is going to be achieved.

These strategies draw heavily on the factors that affect ICT adoption and the factors that affect the success of the ICT adopted. The literature reviewed for this study has identified a number of factors affecting the success of ICT in SMEs in the information systems domain (c.f. Chapter 5, section 5.3.4.). After the initial analysis of the studies containing these factors, a number of items were collected from the works of various authors and classified according to the nature of the factors, namely organisational characteristics, organisational action, system characteristics, internal expertise and external expertise. These categories are based on a classification proposed by Yap et al. (1992), but include their own findings and those of a number of subsequent studies. These items and their sources are detailed in Table 5.4.

In the present work, these factors are represented in the left part of the model (c.f. Figure 6.3) and were grouped into the characteristics of the freelance translator, the characteristics of the freelance translation business, and the translators' perceptions of specialised ICT. The determinants of ICT adoption are summarised in Table 6.7.

6.3.3 Impacts of specialised ICT adoption

The outcome of adopting specialised ICT has direct consequences on the MIPs' performance as a micro business, which can be translated into benefits realised from the

Table 6.7 **Determinants of ICT adoption**

Translator characteristics	Source
1. CEO IT knowledge	DeLone, 1988; Palvia et al., 1994
2. Internal IS/ICT competencies	Caldeira and Ward, 2002
3. User computer literacy	Montazemi, 1988; Igbaria et al., 1998
4. CBIS experience	Raymond, 1985; Igbaria et al., 1998
5. CEO involvement	Cragg and King, 1993
6. Managerial time	Baker, 1987; Cragg and King, 1993; King and McAulay, 1989; Lefebvre and Lefebvre, 1988
Translation business characteristics	**Source**
7. Computer planning	DeLone, 1988
8. In-house processing	Raymond, 1985
9. Financial resources	Caldeira and Ward, 2002; Cragg and King, 1993
10. Age of company	Palvia et al., 1994
Perceptions of specialised ICT	**Source**
11. CEO support and attitude towards IS/ICT adoption and use	DeLone, 1988; Caldeira and Ward, 2002; Thong et al., 1996
12. User attitudes	Caldeira and Ward, 2002

adoption of specialised ICT and problems that this adoption may have originated. There is a need to define the factors that lead to a successful adoption and to benefits for the business, and to identify problems that are likely to arise in the process of adopting ICT or when reaching a higher degree of sophistication.

This study focuses on the subjective measures of organisational performance and adopted the instrument developed by Khandwalla (1977) to measure the index of subjective performance based on the manager's assessment of the company's ability relative to its competitors. While the instrument was originally developed and tested in large organisations, it has also been adapted and validated in the SMEs context (Miller and Droge, 1986; Raymond et al., 1995; Hussin et al., 2002; Ismail, 2004). This resulted in five items measuring MIPs' performance, which are represented on the right-hand side of the model, and summarised in Table 6.8.

Table 6.8 **Impacts of specialised ICT adoption (Khandwalla, 1977)**

Items
1. Long-term profitability
2. Amount of translation work undertaken
3. Financial resources (liquidity and investment capacity)
4. Client base
5. Professional image and client loyalty

6.3.4 Strengths and limitations of the research model

The proposed research model follows an interdisciplinary approach in order to devise a model on ICT adoption with a focus on specialist ICT in a particular sector (multilingual communication), in the context of micro businesses (multilingual information professionals). Information Systems research on models of IT adoption has been used with the objective of testing its findings in this specific context. This can be considered as an innovation in the Translation Studies domain, since previous models focused on purely linguistic aspects of the translation businesses and have not addressed the particular context of the freelance translator.

One of the limitations observed in most of the ICT adoption stage models available, that is, the impossibility of considering future technologies, has been taken into account in the development of the model. Most ICT adoption models (for example, Churchill et al., 1969; Nolan, 1973; Nolan, 1979) describe the evolution of IS assuming certain technologies as driving forces (Saarinen, 1989). This is also the case of previous research in Translation Studies, with models classifying translation technologies by the linguistic processes involved, or by types of application (for example, Melby, 1998; Austermühl, 2001) which have been taken into account. While they focused on previous or current types of applications, the proposed model identifies a core specialist technology (i.e. CAT tools), which is part of a bigger picture of ICT available to MIP that supports the activities that are part of the MIP's workflow. For this reason, this research model presents a comprehensive framework for the use of ICT by MIPs that studies the use of CAT tools in the context of the ICT supporting MIP activities.

A preliminary limitation observed is the potential difficulty in measuring the impacts of specialised ICT in translation businesses, as there are no established measures in the Translation Studies or the IS domain that can be straightforwardly applied in the context of MIPs. Previous research looking at CAT tools has claimed a number of benefits arising from the use of these tools (such as an increased volume of work in less time, or improvements in the cohesion and coherence of translations); however, few empirical studies have directly reported the benefits of using CAT tools in translation businesses, and even fewer have provided evidence of such benefits in the freelance translator context. The next two chapters present a research method for studying these phenomena, i.e. how to apply the research model through the use of valid research methods, and the empirical results obtained by previous research of the author to validate the overall framework, model and measures defined here.

Research methods for studying multilingual information management: an empirical investigation

"Cross-disciplinary research, [...] requires familiarity with measuring techniques in more than one discipline."
(Oppenheim, 1992, p. 7)

In order to illustrate how the presented research framework and model can be applied to investigate a community of practitioners, this chapter presents the research method followed during previous work undertaken by the author between 2002 and 2005.[1] Therefore, this chapter is focused on presenting a suitable research method to design instruments based on previous validated research that enable enquiring about the aspects identified in the research model, and on presenting a plan to analyse the data through quantitative and qualitative techniques. To do so, first some of the research methods used by previous studies in related research areas are reviewed, notably in ICT adoption research and in translation studies research, and a specific research design is proposed and justified. Next, the design of the research instruments and data collection methods used are detailed. Finally, an overview of the data analysis approach performed is presented.

7.1 Research approaches

Studies in the disciplines informing this interdisciplinary research have followed various research approaches, which are summarised in this section.

Easterby-Smith et al. (1999) identify three reasons that the exploration of philosophy may be significant with regard to research methodology: first, it can help the researcher to refine and specify the research methods to be used in a study, that is, to clarify the overall research strategy to be used. This would include the type of evidence gathered and its origin, the way in which such evidence is interpreted, and how it helps to answer the research questions posed. Second, a knowledge of research philosophy will enable and assist the researcher to evaluate different methods and avoid inappropriate use and unnecessary work by identifying the limitations of particular

---

[1] The research was part of a research project funded by the Engineering and Physical Sciences Research Council (EPSRC) of the United Kingdom during the period 2002-2005. The project was entitled *The adoption of translation software by translation SMEs: a study of productivity and organisational issues* (code GR/R71795/01) and was led by Dr. Heather Fulford at Loughborough University, United Kingdom.

approaches at an early stage. Third, it may help the researcher to be creative and innovative in either the selection or adaptation of methods that were previously outside his or her experience. There has long been a debate about the underlying philosophy that should guide valid research, where quantitative research has generally been associated with positivist philosophical traditions, and qualitative research has commonly been associated with interpretivism. Within the social sciences there is an increasing belief that this debate between positivists and anti-positivists is inadequate to address the problems facing researchers in today's world (Cornford and Smithson, 1996).

Oppenheim suggested that "no single approach is always or necessarily superior; it all depends on what we need to find out and on the type of question to which we seek an answer" (Oppenheim, 1992, p. 12). It is now widely acknowledged that a research approach (or strategy) must be selected according to the object of study, the specific research questions and objectives of that study, and the setting in which the research is undertaken so that valid answers are obtained to the research questions. This choice should then be more a matter of appropriateness, rather than a decision only driven by a philosophical perspective (Crossan, 2003).

A number of research approaches are available to information systems researchers, each having its own strengths and weaknesses (Mumford et al., 1985). Galliers (1992) proposed a taxonomy[2] of information systems research approaches which identified and compared the following ten research strategies used in IS research: theorem proof, laboratory experiments, field experiments, case studies, surveys, forecasting, simulation, argumentative studies, interpretive studies and action research.

Although various research strategies have been employed to study ICT adoption in small firms, including interview-based studies (Baker, 1987), longitudinal studies (Cragg and King, 1993), and case studies (King and McAulay, 1989; Caldeira and Ward, 2002), most of the previous research undertaken in this area by the end of the 20th century had been conducted from a positivist perspective[3] through the use of questionnaire-based surveys (see, for example, Raymond, 1985; Lees, 1987; Raymond, 1987; DeLone, 1988; Lefebvre and Lefebvre, 1988; Montazemi, 1988; Raymond, 1989; Kagan et al., 1990; Yap et al., 1992; Daniel et al., 2002).

Much of the research on ICT adoption in the translation sector has been devoted to the adoption of translation tools. However, previous studies have not tended to be focused specifically on freelance translation businesses, but rather considered translation tools adoption in translation businesses in general, in large organisations, or a mix of both large and small translation organisations. Among these studies there have only been a few in-depth surveys (see, for example, Webb, 2000; Fulford, 2001; Höcker, 2003; Lommel, 2004; Fulford and Granell-Zafra, 2005; Dillon and Fraser, 2006; Lagoudaki, 2006), and a number of case studies on translation tools adoption in large organisations (King, 1998; Jaekel, 2000; Lange and Bennett, 2000). Most of the studies in the translation sector, like in the area of IS, have been conducted from a positivist perspective.

---

[2] This taxonomy is a revised and amended version of the one proposed in Galliers and Land (1987), and then published in Galliers (1992).

[3] Orlikowski and Baroudi (1991) noted that 96.8% of research in leading information systems journals followed the positivist tradition.

Interpretivist approaches (i.e. simulation, argumentative studies, interpretive studies and action research), and the other empirical approaches available (i.e. theorem proof, laboratory experiments, field experiments and forecasting), according to Galliers' taxonomy (1992), have not been used much in previous studies of ICT adoption, and did not seem to suit the particular context of this research. Accordingly, survey and case study research approaches are further discussed below.

Survey research permits the examination of a phenomenon in a wide variety of natural settings. This examination essentially comprises a snapshot of practices, situations or views at a particular point in time, and is typically undertaken using questionnaires or structured interviews, from which inferences may be drawn (Galliers, 1992). In survey-based research the researcher has a clearly defined model with independent and dependent variables and the factors that affect it, so that anticipated relationships can be tested against observations of the phenomenon under investigation. The major strength of the survey approach is that it permits the collection of data from a large number of subjects, thus allowing quantitative analysis to test inferences, and giving the potential to generalise the findings to an even larger number of cases. One of the major disadvantages of this approach is that the variables under study have to be known in advance. Thus, it can only be used in relatively well understood situations.

The case study approach is commonly used in the study of issues that are not well understood, or where relationships between the context and the phenomenon under study are not clear (Yin and Campbell, 2002, p. 13). The strength of the case study approach is that it enables the researcher to capture reality in considerably greater detail than is possible in a questionnaire survey. Weaknesses of the case study approach include the fact that its application is usually restricted to a single organisation, or to just a small number of organisations.

7.2 Selecting a suitable approach

In order to investigate the adoption of ICT by freelance translators in the UK and the factors affecting that adoption, a survey approach was deemed to be the most suitable method for data collection. A key advantage of the this approach is that it permits the collection of data from a large number of subjects and thus enables gathering data from a larger sample of professionals to obtain an overall picture of the freelancers of the multilingual information management sector in the UK. The types of survey frequently used in social research include questionnaires, interview, observation studies, and content analyses (Bryman and Bell, 2003). For practical reasons, a questionnaire survey was deemed to be a suitable method for data collection in the present study, since the researcher had access to lists of professional translators that belonged to a professional association. Therefore, the mailed questionnaire method seemed the most suitable way for obtaining a substantial number of responses from a geographically dispersed sample. The options of using telephone or email surveys were discarded as suitable contact details were not available to the researcher and in an open online survey there would have been no straightforward means of control over the reliability of the eventual participants.

The survey approach was deemed to be more powerful than a case study approach for quantifying relationships in the research model. However, questionnaire-based surveys can be weak at providing insights about particular issues. Therefore, a decision was made to follow the gathering of quantitative data in the questionnaire survey with a more in-depth study of individual cases in order to permit a closer examination of issues and relationships. The initial survey exploring the adoption of ICT by translators in the UK was then followed by further fieldwork in order to gain more detailed insights about the decision to adopt CAT tools of the participants. The aim of this second phase of the fieldwork was to shift the study from a "descriptive" to an "analytical" design (Oppenheim, 1992, p. 12).

A number of research design options were considered for this subsequent phase of the research, including interviews, focus groups and a further questionnaire survey. Focus groups are typically designed to collect data from a limited number of participants through a focused discussion among the group of participants (Lewis-Beck et al., 2003). Such groups may stimulate participants' thinking and elicit ideas about a specific topic (Gliner and Morgan, 2000, p. 341). However, the goals of this second phase of the fieldwork required more detailed information from each individual adopter or non-adopter of specialist ICT, and thus focus groups were rejected. Interviews allow the gathering of data in more detail from each subject, and can be conducted either face-to-face, by telephone or by online methods. However, a large number of interviews would have had a high economic cost and they would have required a longer timescale than the available one, so interviews were ruled out because they did not allow the collection of data from a large number of translators.

In the light of the above drawbacks of interviews and focus groups, a decision was made to conduct a further questionnaire-based survey. A survey approach would again allow collecting data from a geographically dispersed and larger sample without incurring greater expenses and, at the same time, it would allow us to ask open-ended questions to capture qualitative information. In addition, since the participants were identified from the sample of the first survey, had given their consent to collaborate further, and had facilitated valid email addresses, an online survey was deemed a suitable method to conduct this phase of the research.

Online data collection methods are regarded as a faster and less costly way of collecting a larger volume of detailed information in shorter time scales than traditional data collection methods (Gaiser, 1997; Clarke, 2000; Dillman, 2000). At the same time, these methods simplify the data collection process, allow the collection of a large number of responses, and can ensure the accuracy of the collected data by adding responses automatically into a database, offering significant savings in terms of data entry (Mann and Stewart, 2000). In contrast with traditional data collection methods, some extra time should be devoted to designing and planning this type of online study, but once ready, the time required to collect the data and prepare the responses for their analysis would be reduced.

The online survey included a combination of a web-page-based questionnaire with email messages similar to email surveys (Mann and Stewart, 2000), thereby gaining some of the advantages of this type of survey, too. For example, the convenience of initially contacting participants to ask them to collaborate further in this research, to explain to them how to access the website containing the questionnaire, and to open a communication channel with the researcher should they have any queries. The advantages and potential of an online survey were clear in terms of gaining access to the participants quickly, cost and time savings, eliminating transcription bias, and easier handling of data than by manual methods (for a more detailed discussion of the benefits of online research methods, see Mann and Stewart, 2000).

Nevertheless, such methods also have disadvantages that can represent a challenge for the researcher (also discussed in detail by Mann and Stewart). In fact, these authors claim that "[o]nce a researcher has decided that CMC [Computer-Mediated Communication] is the way forward, the first practical step is to ensure that the researcher and all respondents have access to the required technology and the confidence to use it" (Mann and Stewart, 2000, p. 66). The first challenge, therefore, relates to the technological infrastructure and knowledge required to set up the survey and get the responses from the participants. Also, "establishing contact for individual person-to-person CMC usually involves a mutual exchange of email addresses" (ibid), which sometimes can create problems when recruiting participants and contacting them without their consent, which could be considered "spamming."

These initial challenges were overcome by the characteristics of the sample and the researcher. The sample consisted of a number of participants from the first survey who (1) had previously agreed to collaborate further with this research after participating, (2) provided their email addresses among their contact details, and (3) were regular email users and familiar with Internet technologies. The researcher's expertise in HTML programming and Internet technologies was satisfactory to set up the study and run it to the expected standards.[4]

The use of mixed approaches is often recommended in the literature. Bryman maintained that "the use of more than one method [...] can greatly enhance the process of fusing problem and method, by allowing the researcher to reap the opportunities presented by two or more techniques" (Bryman, 1989, p. 248). Others point out the legitimacy of combining methods in the research design if such a strategy addresses the study's research questions (Miles and Huberman, 1994; Brannen, 1995). Burgess, (1984) defines "multiple research strategies" as the use of diverse methods in tackling a research problem. This strategy has traditionally been referred to as "triangulation" (Denzin, 1970). Denzin distinguishes four different types of triangulation: multiple methods, multiple investigators, multiple data sets and multiple theories. Multiple methods triangulation can be between-methods or within-method. A within-method approach involves the same method being used on different occasions, while between-methods means using different methods in relation to the same object of study. The design of this study fits into the "multiple methods"–"between methods" approach defined by Denzin and was structured in two phases:

- **Phase 1: ICT adoption and use exploratory study**; conducted through a mailed questionnaire;
- **Phase 2: In-depth analysis of organisational impacts and evaluation of ICT sophistication**; conducted through online survey methods.

---

[4] This taxonomy is a revised and amended version of the one proposed in Galliers and Land (1987), and then published in Galliers (1992).

7.3 How to explore ICT adoption and use

As explained before, the exploratory study comprised a mailed questionnaire survey to UK freelance translators. Its main goal was to obtain a snapshot of ICT usage among freelance translators, that is, to provide a profile of translators and their use of general-purpose ICT and translation-specialised tools and resources, as well as their attitudes towards them. In addition, this questionnaire survey permitted the identification of a sample of CAT tools users and CAT tool non-users that could be studied further in the next phase of the research.

7.3.1 Questionnaire design considerations

The questionnaire was designed following suggestions from previous studies about survey methods, mainly Dillman's recommendations for a total (1978) and tailored (2000) design method. Dillman's Total Design Method (TDM) for the development and use of a mail questionnaire (1978) was borne in mind in order to ensure a satisfactory response rate following the maxims of minimising the cost for responding, maximising the rewards for doing so, and establishing trust that those rewards would be delivered. Dillman's most recent contribution to the design of mail surveys (defined as the Tailored Design Method in Dillman, 2000) also provided useful recommendations on writing questions, constructing the questionnaire and implementing the survey.

Following Dillman's recommendations, a cover letter was produced along with the questionnaire. This letter highlighted the relevance of the study, the importance of the participation of the translators, as well as the benefits that they would obtain from participating in the study. As an additional incentive to encourage participation in the study and increase the response rate, two measures that were adopted: the offer to respondents of a copy of a summary of the findings of the study and an additional prize draw with prizes of book vouchers for three of the respondents. Following another recommendation made in the survey research literature, stamped addressed envelopes were provided for return of the questionnaire. In addition to the cover letter, a letter of endorsement was enclosed with the questionnaire. This letter was written by the Director and Chief Executive of the professional body from which the mailing list was obtained.

All these measures contributed to delivering a reliable questionnaire that was clear, interesting, easy to return and had a professional layout. In addition, these efforts would help to involve translators in the study and increase the response rate substantially.

The appearance of the questionnaire was kept simple and visually attractive. Sans serif fonts were used in the questionnaire to increase its readability, as well as bold format and text boxes to enable the identification of the different sections of the questionnaire. The front cover was also simple, but eye-catching, and it clearly identified the researcher's name, organisation and contact information, as well as a title that was directly related to the study. Moreover, basic instructions for completing the questionnaire, as well as a clause ensuring confidentiality and thanking the participants for their cooperation, were included at the bottom of the front cover. A copy of the questionnaire is included in Appendix 1.

7.3.2 Instruments and structure of the questionnaire

In order to establish the sequence of the sections of the questionnaire, Dillman's suggestions were considered for increasing the respondents' motivation for, and confidence in, completing the questionnaire (Dillman, 1978). Accordingly, questions that were similar in content or type were grouped together. The order of the questions took advantage of the cognitive ties that respondents were likely to make among the groups of questions. Questions that were more likely to be difficult were placed after questions that were likely to be easier to answer.

As a result, the questionnaire was structured in four sections as follows:

- Section A: Translator profile
- Section B: Information Technology (IT) usage
- Section C: Internet usage
- Section D: IT strategy

Where appropriate, validated research instruments were drawn from existing research on ICT adoption and adapted for this study. The sections of the questionnaire and the instruments used are now presented in detail.

7.3.2.1 Section A: Translator profile

This section contained questions regarding translators' characteristics and the characteristics of their translation business. In particular the following issues were addressed:

- *educational background,* such as educational level and translation qualifications;
- *demographic data,* such as age and gender;
- *data about their employment situation,* such as whether they were working in-house, as a freelancer, or managing a translation company; how many years they had been working as translators; their workload; the services they provided; the language combinations they translated; and the subject fields they worked in.

Also, in order to help with the pre-screening of freelance translators, translators were required, immediately after the section on translator profile, not to continue with the rest of the questionnaire if they were not, at that point in time, actively involved in translation work.

7.3.2.2 Section B: Information Technology usage

This section consisted of questions about the IT skills and IT qualifications of the translators, followed by a question about their familiarity with, and their working knowledge of, a number of software applications.

With regard to IT current usage, Raymond and Paré's instrument for measuring IT sophistication in small manufacturing businesses (Raymond and Paré, 1992) was used to develop the question about IT usage (Question 17). These authors defined IT usage in terms of "technological sophistication" and "informational sophistication." These aspects of IT sophistication basically refer to "the number or diversity of information technologies used by small businesses" (i.e. the type of technology used) and "the nature of the applications portfolio" (i.e. the function of the type of applications). In this study, the translators were asked to indicate whether they were currently using a number of IT applications, which had been grouped according to their function (e.g. document production, business management, translation production).

With regard to IT knowledge, Magal and Lewis (1995, p. 76) defined IT knowledge "in terms of awareness of, familiarity with, exposure to, or a working knowledge of technology, rather than expertise." They measured IT awareness by providing a representative list of software commonly used by SMEs and asked the respondent to indicate the extent to which they were familiar with the application or software. Hussin et al. (2002) and Ismail (2004) adapted and tested the measures with a sample of small manufacturing firms in the UK and with a sample of Malaysian manufacturing SMEs, respectively. The survey also adapted Magal and Lewis's instrument to design the question about IT usage. The respondents were asked to indicate their level of familiarity with, and knowledge of, IT on a 4-point Likert-type scale with the following (ordinal) categories: a score of 1 represented "Not familiar," a score of 2 represented "Familiar, but with no working knowledge," a score of 3 represented "Familiar, with some working knowledge," and a score of 4 represented "Familiar, with extensive working knowledge."

7.3.2.3 Section C: Internet usage

This section of the questionnaire was designed similarly to the second part of section B, but in this case translators were asked about their familiarity and working knowledge with Internet-based tools and resources. It was separated from the earlier question on IT usage for pragmatic reasons of length and format and to explicitly ask translators about the online information sources and ways of accessing knowledge they used. Participants were also asked whether they had their own website to promote their services.

For the question on usage and experience with Internet-based tools and resources (Question 19), Raymond and Paré's (1992) and Magal and Lewis' (1995) instruments were used in a similar way to that of the earlier question of usage. Similarly, the same 4-point Likert-type scale was used for this purpose.

7.3.2.4 Section D: IT strategy

This section of the questionnaire consisted of three different types of questions regarding translators' opinions on the technologies referred to in sections B and C of the questionnaire. After asking if the translators had a written business plan, a question was asked about the translators' opinions on the importance and current use of IT for a number of tasks undertaken by translators. Chenhall and Morris (1986) developed and tested an instrument designed to measure accounting information systems (AIS) design in large organisational context, which was later adapted and tested in the small business context by Gul (1991), and more recently to measure both AIS requirement and AIS capacity in accounting SMEs by Ismail (2004). The scales used by Ismail's instrument were used in the study to obtain translators' opinions on the relevance of

IT in their work. This relevance and use were measured using a 4-point scale ranging from "Not important" to "Very important" and from "None" to "Extensive" use.

Questions 22 and 23 measured the perceptions that translators have of ICT in general, and CAT tools in particular. The instruments used in these questions were based on one developed by Cragg (1990) to measure computerisation success, where the author was asking the respondents what they thought of a number of issues regarding the use of computers. In Question 22, the original instrument measured the attitudes of users and non-users of computers, which were phrased accordingly to obtain the information on translators' perceptions of ICT in general. In Question 23, the statements in the original instrument on ICT in general were worded differently to capture translators' perceptions of one particular type of ICT (CAT tools). One of the items ("Computerisation significantly improves my communication with customers") could not apply to the use of CAT tools, and was substituted by an item asking about the respondents' opinion on the cost of these tools (i.e. "CAT tools are well worth their cost"), which was one of the translators' concerns that arose from the literature examined on CAT tools.

The last question (Question 24) further enquired about the translators' ICT strategy. This question was adapted from Hussin's instrument (1998) to measure various aspects of ICT strategy, which covered Earl's (1989) three levels of strategy related to ICT, namely, the Information Systems strategy, the Information Management strategy, and the Information Technology strategy. Respondents were asked to indicate their position along a 5-point scale with regard to the bipolar alternatives listed in Figure 7.1.

Finally, the back cover of the questionnaire included a free text box for respondents to comment on the questionnaire or the study as a whole. A question was also included about whether the participant would be willing to participate in other stages of the research. Also, there was space for providing the participant's contact details (name,

A						B
I treat each decision about a new IT investment independently	1	2	3	4	5	My decisions about IT investments are guided by a formal IT strategy
I am concerned with using IT to solve short-term problems	1	2	3	4	5	I am concerned with using IT to solve medium to longer-term problems
I am concerned with matching technology to my business needs	1	2	3	4	5	I am concerned with getting the most up-to-date technology
I am concerned with how to better manage my IT resources	1	2	3	4	5	Managing IT is not as critical as managing other non-translation related resources
I am concerned with achieving a greater level of integration of my computer systems	1	2	3	4	5	I am concerned that the majority of my computer systems remain as standalone applications
The primary benefits I seek from IT are improved productivity and efficiency	1	2	3	4	5	Computer systems bring a wide range of benefits including competitive advantage

Figure 7.1 IT strategy bipolar items

address, and email) if they wanted to receive a copy of the summary of the survey findings.

7.3.3 Questionnaire refinement

As suggested by Dillman (1978) the questionnaire was refined before carrying out the data collection. The refinement of the questionnaire followed two stages: pre-testing by academics and research students and pre-testing by translators.

The questionnaire was first pre-tested by academics in the researcher's university department, who had prior experience with surveys on ICT adoption, and also by fellow research students, who had designed and used questionnaires in their research. Useful feedback arose from these pre-tests. Overall, the layout of the questionnaire was perceived to be appealing, neat and easy to follow, although comments from the pre-testers helped with the rewording of some questions and statements to clarify their meaning, with the reordering of some statements to make them easier to follow, and with improvements to the layout. Based on this feedback, amendments were made to the questionnaire.

After pre-testing the questionnaire in an academic environment, it was rigorously pre-tested by six translators who were likely to have similar characteristics to the participants in the survey. This rigorous process of assessing the questionnaire's content and clarity provided useful feedback which was used to improve it, until the questions could be clearly understood and answered without problems by the translators. The translators were also asked to answer the questionnaire and the overall impression from the answers given and the comments made by them was that the questionnaire was appropriate and ready to be delivered to the translators in the sample.

7.4 How to analyse organisational impacts and evaluate ICT sophistication

As in the first phase of the study, the design of the online study involved the development of the survey, adapting existing scales and measures where available, managing the technical issues involved in the creation of the web form, testing and piloting of the survey, the preparation of contact messages, handling the responses received, and keeping track of the participants' interaction and their responses to the survey. These issues are explained in more detail in this section. The questionnaire for the online survey was designed similarly to a semi-structured questionnaire using open-ended questions to obtain both quantitative and qualitative information from the respondents, as opposed to the rigidity of using a structured questionnaire. Therefore, as noted by Easterby-Easterby-Smith et al. (1999, p. 112), this design is suitable for being analysed following the method suggested by Miles and Huberman (1994). Briefly, the analysis of the data using this method is accomplished by drawing conclusions from the visual patterns observed in a matrix sheet that displays the data extracted from the responses through a data reduction process (e.g. coding). This method is further explained in the section of this chapter presenting the analysis of the research.

7.4.1 Online questionnaire design considerations

A number of issues affecting online research methods were considered for the design of the online study. This section focuses on how such issues affected this study and how they were addressed.

In order to ensure that the participants had access to the required technology and the confidence to use it, an online study should be kept as easy as possible for respondents to access and complete (Dillman, 2000). Also, the virtual environment of the survey should be familiar to the respondents (Mann and Stewart, 2000). In seeking these priorities, some advantages from email and web-page-based surveys were met. First, all the communications between the researcher and the participants were made through text-based email messages, making them convenient for the respondents because they required no facilities or expertise beyond those that they use in their day-to-day email communication.[5] Second, a website containing a form was used to collect the data from the survey, avoiding typical problems of email-based surveys, such as selecting several answers when only one choice is required, deleting questions accidentally, or altering their format (Mann and Stewart, 2000), and providing a visually attractive interface[6] that appeared identical to all respondents, was easy to complete and submit, and whose data was in a completely predictable and consistent (coded) format, making automated processing and analysis possible by the researcher.[7]

Another challenge may arise from the perception of the notification email as from an unknown sender by the participants (Faught et al., 2004). Although participants were familiar with this research and had previously agreed to collaborate further, previous communication with them was accomplished by postal mail. Therefore, it was possible that the first contact email informing the participants about the second phase of the fieldwork seemed unknown to them, their email clients filtered the message as "junk"/"spam" mail, or that they simply deleted the message before reading it. These problems are analogous to the "wastebasket problem" for mail surveys, and the researcher needed to be aware of this issue and work to avoid filters and the delete button. To overcome this potential problem the researcher's university email system was used, which identified the sender's email address belonging to a UK university, and also used the university's mail server, increasing the reliability of message handling. Moreover, neither graphic elements nor attachments were sent along with the messages and email messages intentionally contained only text-based information, therefore reducing the risk of some email clients blocking the message for being potentially dangerous.

---

[5] Other options such as online discussion groups were also considered, but email communication avoided the need for login procedures, preserved the privacy of the messages between the participants and the researcher, and allowed contacting each participant in a personal way. Moreover, since no discussion among participants was required, email messaging covered the communication requirements for this study satisfactorily.

[6] As opposed to long and dull email messages using plain text without any format typically employed in email surveys.

[7] Text-based email surveys, once again, require additional editing before processing the responses.

7.4.1.1 An online survey

Prior to the development of the web form, and to sending contact messages to the participants, a survey implementation strategy was devised.

First, a tracking document was created using spreadsheet software and included information about the participants, the messages sent to them, the messages received from them, and the overall progress of the survey.

Second, contact messages to be sent were prepared and produced using group mail software.[8] These messages included an invitation letter to participate in the online study, a thank you message for completing the survey (to be sent individually or in small groups after receiving the responses), and a template for an apology message in the event of technical problems (to be modified and addressed individually in each case).

Third, participants were split into two groups for sending the contact messages: the pilot group (21), and the rest of the participants (130).

After these preliminary tasks, the web page containing the questionnaire was designed. Separate sets of questions were produced for "adopter" and "non-adopter" groups. Once the contents were ready, the web page containing the questionnaire was designed in HTML[9] following guidelines for good web design and principles for constructing and implementing web surveys. In the literature on research methods, it is highlighted that "Internet surveys need to be designed with the less knowledgeable, low-end computer user in mind" (Dillman, 2000, p. 377), and that it is important to design with computer and questionnaire logic in mind: "Meshing the demands of questionnaire logic and computer logic creates a need for instructions and assistance, which can easily be overlooked by the designer who takes for granted the respondent's facility with computer and web software. [...] The building of such instructions takes on the same level of importance as writing the survey questions" (idem).

Some of the design principles for web surveys discussed by Dillman (idem) and applied to this web survey were addressed as follows:

- "Introduce the Web questionnaire with a welcome screen that is motivational, emphasizes the ease of responding, and instructs respondents about how to proceed to the next page" (p. 377). This was achieved by including welcoming, motivation, and instructions statements in the contact email and also contained a link pointing to the web address where the web survey was located.
- "Provide a PIN number for limiting access only to people in the sample" (p. 378). Since the invitation to participate was sent privately to the participants, password protection access was considered unnecessary; moreover, for ease of use reasons, it was considered that a login process could complicate the access to the survey or deter some respondents from following this process.
- "Present each question in a conventional format similar to that normally used on paper self-administered questionnaires" (p. 379). Questions and scales were reproduced in a similar

---

[8] After considering several choices of multiple emailing software applications, *Infacta Group Mail free edition* (version 3.4.206) was used to create groups of contacts, create the messages and send the messages to each group of participants in one go, but individually addressed.

[9] HTML stands for HyperText Markup Language, the coding system used to create pages which can be displayed by web browsers.

way to a paper questionnaire, although taking advantage of colouring, layout, and shading features offered by HTML format.
- "Restrain use of color so that figure/ground consistency and readability are maintained, navigational flow is unimpeded, and measurement properties of questions are maintained" (p. 382). As mentioned above, HTML format features were used, but only using colouring and bold font face in a sensible way, so that it enhanced the readability of the questions. Moreover, all the questions including a list of items across scales were designed using resizable tables to ensure the integrity of the proportions and consistency were maintained.
- "Avoid differences in the visual appearance of questions that result from different screen configurations, operating systems, browsers, partial screen displays, and wrap-around text" (p. 385). All text used relative font sizes so text could be enlarged or reduced using the text size options available in visual browsers, and a flexible page format was used so pages could be automatically resized for different window sizes and screen resolutions avoiding annoying wrap-around effects regardless of the participants' computer or software used to display the web survey.
- "Provide skip directions in a way that encourages marking of answers and being able to click to the next applicable question" (p. 394). This principle was applied several times giving explicit instructions to click on a link that forwarded the respondent to the following question (i.e. "please click here to go to next section").

The final version of the survey web page is reproduced in Appendix 2, and was made available online for the duration of the study through the university web servers.

7.4.2 Instruments and structure of the online questionnaire

As in the design of the questionnaire in the first phase of the study, suggestions made in research methods' literature were considered to define the structure and sequence of the sections in the online questionnaire (Dillman, 2000). Accordingly, questions that were similar in content or question type were grouped together. The order of the questions took advantage of the cognitive ties that respondents were likely to make among the groups of questions. Questions that were more likely to be difficult were placed after questions that were likely to be easier to answer.

The web questionnaire was structured in four parts (A-D):

- Part A: Terminology management tools
- Part B: Translation memory
- Part C: Your 'translation toolkit'
- Part D: Your profile

Standard instruments were used for the survey questions where possible, drawing on existing instruments used in previous research on ICT adoption and adapting them for this study. The main instruments used are presented in detail below, and a complete copy of the questionnaires developed for adopters and non-adopters of CAT tools can be found in Appendices 2 and 3.

7.4.2.1 CAT tools: terminology management tools and translation memory

Questions in Parts A and B of the study were formulated using Moore and Benbasat's instrument, which was designed to "measure the various perceptions that an individual may have of adopting an information technology (IT) innovation" (Moore and

Benbasat, 1991, p. 192). Their instrument was designed to "be generally applicable to a wide variety of innovations, especially other types of information technologies" (p. 194). They developed this instrument using the theoretical framework of innovation adoption developed by Rogers (1995). In this instrument, eight constructs were used to measure the perceptions of adopting an information technology innovation. According to Moore and Benbasat, the reason for using the *perceived* characteristics of innovations, rather than perceptions of the innovation itself, was that "the findings of many studies which have examined the primary characteristics of innovations have been inconsistent" (p. 194). They argue that "primary attributes are intrinsic to an innovation independent of their perception by potential adopters," while "the behaviour of individuals [...] is predicated by how they perceive these primary attributes" (p. 194). In addition, they further claim that "studying the interaction among the perceived attributes of innovations helps the establishment of a general theory" (p. 194).

The constructs used in the instrument were relative advantage, compatibility, voluntariness, image, ease of use, result demonstrability, visibility, and trialability. These were defined by Moore and Benbasat as follows:

- *relative advantage:* "the degree to which using an innovation is perceived as being better than its precursor" (p. 196);
- *compatibility:* "the degree to which [using] an innovation is perceived as being consistent with the existing values, needs, and past experiences of potential adopters" (p. 195);
- *voluntariness:* "the degree to which use of the innovation is perceived as being voluntary, or of free will" (p. 195);
- *iImage:* "the degree to which use of an innovation is perceived to enhance one's image or status in one's social system" (p. 195);
- *ease of use:* "the degree to which an individual believes that using a particular system would be free of physical and mental effort" (p. 197, cited from Davis, 1986, p. 82);
- *result demonstrability:* the degree to which "the results of using [an] innovation" are communicated (p. 203);
- *visibility:* the degree to which "the results of using [an] innovation" can be observed (p. 203);
- *trialability:* "the degree to which an innovation may be experienced with before adoption" (p. 195).

The in-depth study of the second phase of the research required an instrument that measured the adoption of CAT tools (i.e. an IT innovation) within freelance translation businesses (i.e. a very small organisation). Furthermore, Moore and Benbasat's instrument was measuring the perceptions of an individual, which perfectly suited the case of the translation micro businesses under study, where the freelance translators were not only the users of the technology, but also the ones making the decision of adopting CAT tools.

Questions 1 to 3 of parts A and B were based on Moore and Benbasat's instrument. All the items of the constructs in the complete instrument that were applicable to an individual were used in these questions. The wording was slightly modified to reflect the purpose of this research and the specific context of freelance translators. In addition, the same items were used for the version of the questionnaire addressed to non-adopters of CAT tools, accordingly rewording them by using conditional verb tenses that enabled obtaining information about their opinion on how using the tools would influence their work (e.g. instead of asking them if translation memory *enables* them to accomplish tasks more quickly, the statement read if translation memory *would enable* them to accomplish tasks more quickly). All perceptual items in the instrument were measured by five-point Likert scales representing a range from "Strongly Disagree" to "Strongly Agree." The items for each question and the constructs they relate to are presented in Tables 7.1 to 7.4.

Table 7.1 Items for Question 1: using terminology management tools/translation memory

	Item	Construct
(In Part A) Terminology management tools...	1. Enable me to accomplish tasks more quickly. 2. Improve the quality of work I do. 3. Make it easier for me to do my job. 4. Improve my job performance. 5. Are overall advantageous in my job. 6. Enhance my effectiveness in my work. 7. Give me greater control over my work. 8. Increase my productivity.	*RELATIVE ADVANTAGE*
(In Part B) Translation memory...	9. Are compatible with the type of translation assignments I undertake. 10. Fit well with the way I like to work. 11. Are cumbersome to use. 12. Require a lot of mental effort. 13. Are often frustrating. 14. Do what I want to do easily. 15. Are easy for me to use. 16. Were easy for me to learn.	*COMPATIBILITY* *EASE OF USE*

Table 7.2 Items for Question 2: terminology management tools/translation memory and the translation sector

	Item	Construct
(In Part A) Terminology management tools... **(In Part B) Translation memory...**	1. My clients expect me to use them. 2. My use of them is voluntary. 3. Using them improves my image within the translation sector. 4. Clients prefer to work with translators who use them. 5. Translators who use them have a high profile in the translation sector. 6. Having them is a status symbol among translators.	*VOLUNTARINESS* *IMAGE*

Table 7.3 Items for Question 3: learning about terminology management tools/translation memory

	Item	Construct
(In Part A) Terminology management tools…	1. I have seen how other translators use them. 2. Many freelance translators use them. 3. Before deciding whether to use them, I was able to try them out fully. 4. I was permitted to use them on a trial basis long enough to see what they could do. 5. I had ample opportunity to try them out before buying.	*VISIBILITY* *TRIALABILITY*
(In Part B) Translation memory…	6. I would have no difficulty telling others about what they can do. 7. I believe I could communicate to others the advantages and disadvantages of using them. 8. The benefits of using them are apparent to me.	*RESULT DEMONSTRABILITY*

Table 7.4 Items for impacts of terminology management tools/translation memory

	Items
Impacts of Terminology management tools on… // Translation memory on…	Translators' turnover Size of translators' customer base Quality of translators' translations Translators' productivity Volume of work translators undertake Number of clients translators have Volume of work offered to translators by clients Prices translators charge for work they undertake

In Question 4 of Parts A (terminology management tools) and B (translation memory) the participants were asked about the extent to which the use of CAT tools affected a number of elements of the translators' business (for adopters); and the impacts of the potential use of these tools, in the case of CAT tool non-adopters. For this purpose, a 5-point scale ranging from "Large Decrease" to "Large Increase" was used for the items in Table 7.4.

7.4.2.2 Translation business characteristics

In Question 2 of Part D of the online survey, participants were asked to provide some details about their professional background as translators, and the type of work they undertake. For this purpose, questions relating to their translation assignments, the environment of their freelance translation business, and about ways of learning to use new software tools were formulated. In addition, a 5-point performance scale (ranging from very weak to very strong) was adapted to the context of freelance translators from an original performance scale developed by Khandwalla (1997) to measure the index of subjective performance based on the manager's assessment of the company's ability relative to its competitors. While the instrument was originally developed and tested in large organisations, it has also been adapted and validated in the SMEs context (Miller and Droge, 1986; Raymond et al., 1995; Hussin, 1998; Ismail, 2004). This resulted in five items measuring translator's performance, namely through a 5-point scale ranging from "Very weak" to "Very strong": (a) long term profitability, (b) amount of translation work undertaken, (c) financial resources (liquidity and investment capacity), (d) client base, and (e) professional image and client loyalty.

7.4.3 Online survey trial and piloting of the questionnaire

After uploading the web pages containing the survey to the researcher's web space on the university server, a number of access tests were carried out from different locations and using different computer specifications to ensure compatibility with different machines and Internet connections.

Once electronic access to the survey was tested, the questionnaire was piloted to gain some insights into the likely response rate, and the expected types of response, as well as to gain assurance that the wording of the questions was clear. As suggested by Oppenheim (1992), the respondents in pilot studies should be as similar as possible to those in the main survey, so the questionnaire was then sent to 21 random translators for the pilot exercise. Twelve responses were obtained and showed that the questions were clearly understood, that they could be answered without problems by the translators, and the overall impression and the feedback from the piloting respondents was very satisfactory.

7.5 The data analysis scheme

Given the exploratory purpose of the first survey, a quantitative data analysis approach was deemed suitable to draw conclusions that could be statistically generalised to the population under study.[10] On the other hand, the follow-up and in-depth nature of the subsequent survey was aimed at gaining more detailed insights into ICT adoption and consequently, a more qualitative approach to analyse the data was deemed suitable. The following sections briefly describe these two proposals of analysing data, using as an example the analysis undertaken in both phases of the study.

---

[10] The survey questionnaire was mailed to a sample of 1400 translators in the UK; 591 usable responses were received. After a pre-screening exercise, 152 of those responses were eliminated on the grounds that the respondents reported that translation was not their principal job, but rather a subsidiary activity. The remaining 439 valid responses represented a response rate of 35%. Of those 439 valid responses, 48 (11%) were received from in-house translators, and 391 (89%) from freelancers.

7.5.1 A quantitative data analysis approach

Since the purpose of the study used as an example was initially exploratory and the characteristics of the sample under study had a reasonable size, a quantitative approach was taken to analyse data. To do so, descriptive statistics were examined to find out whether the sample was representative or not, and to understand better the study sample. In exploring ICT tool adoption, descriptive statistics, chi-square tests and logistic regression analyses were used to identify levels of ICT tool uptake, and test the relationships between the adoption of CAT tools and the adoption of, and experience with, other ICT that were supporting translators in the activities in their workflow. In order to investigate the characteristics of the freelance translators who had adopted CAT tools and of their freelance translation businesses, chi-square tests and logistic regression analyses were also used, leading to the definition of a profile of CAT tool adopter. Then, factor analysis statistical technique was conducted to explore and examine the perceptions of ICT in general, and CAT tools in particular, among the translators in the survey sample. The perceptions of CAT tool adopters and non-adopters were then compared and their differences assessed through ANOVA tests. In addition, perceptions of the translators with different levels of experience with CAT tools were also compared through ANOVA tests.

A schematic view of this quantitative data analysis plan is depicted in Figure 7.2.

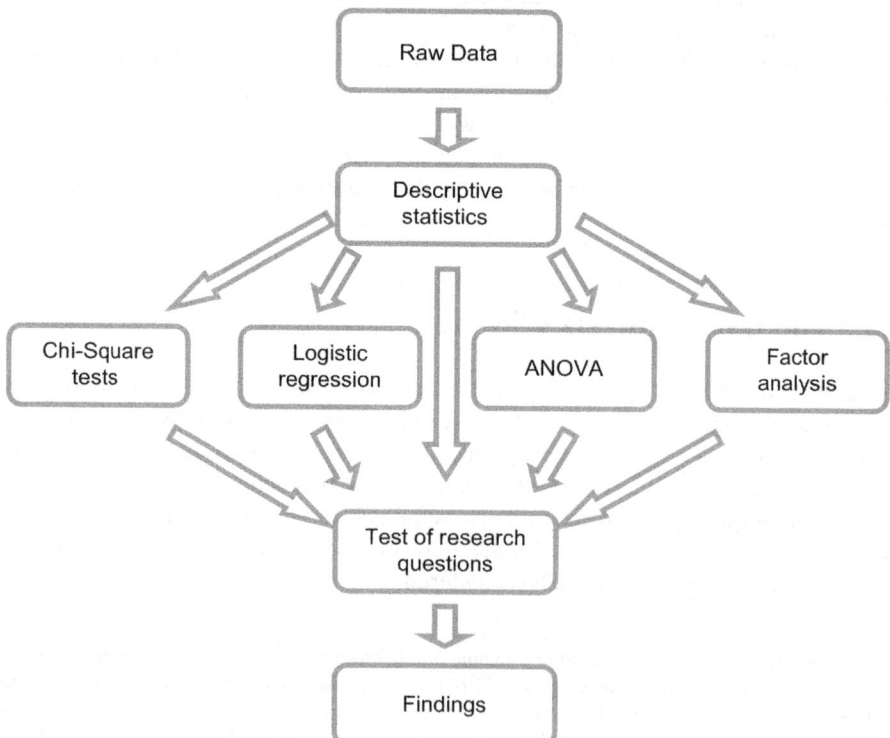

Figure 7.2 Data analysis for an exploratory study

7.5.1.1 Addressing non-response bias and generalisation of results

According to Dillman (2000), there are four possible sources of error in sample survey research. He calls them the "cornerstones for conducting a quality survey" (p. 9). These four sources are sampling error, coverage error, measurement error and non-response error:

- "Sampling error" is the result of surveying only some, and not all, elements of the survey population. This type of error always exists at some level when a random sample is drawn. It can be reduced through larger samples but cannot be eliminated completely unless one conducts a census (Lindner et al., 2001).
- "Coverage error" is the result of not allowing all members of the survey population to have an equal or known nonzero chance of being sampled for population in the survey.
- "Measurement error" is the result of poor question wording or questions being presented in such a way that inaccurate or uninterpretable answers are obtained. Researchers can reduce this type of error by using items that are valid, reliable, and unambiguous to the research subjects (Lindner et al., 2001).
- "Non-response error" is the result of people who respond to a survey being different from sampled individuals who did not respond.

Lambert and Harrington (1990) suggested that potential non-response bias needed to be addressed when response rates fall below 40%, therefore it was important to estimate the effects of non-response bias as it might have affected the generalisability of the survey results.

A variety of ways have been suggested to deal with the potential problem of non-response bias. Amstrong and Overton (1977) and Lindner et al. (2001) discuss extrapolation methods for estimating the response of non-respondents. The extrapolation method is based on the concept that subjects who respond late (either by answering later or by requiring more prodding before answering) have similar characteristics to non-respondents. In this method, known characteristics of groups that respond readily and less readily are compared and extrapolated. If the groups do not differ in their characteristics, it is assumed that there are no systematic differences in their responses, suggesting that the non-response bias is not a significant factor.

In the first survey of this study, the comparison of early to late respondents or "extrapolation method" was adopted to analyse non-response bias. Lindner et al. (2001, p. 52) defined late respondents as those who respond in the last "wave" of respondents in successive follow-ups to a questionnaire. To ensure that the number of late respondents is large enough to be meaningful, both practically and statistically, the respondents were divided into two groups comprising the first 30 and the last 30 responses received. The intermediate respondents were excluded to clearly demarcate early and late respondents. The groups were then compared using some of the main variables measuring the respondents' characteristics and usage of ICT using the Mann-Whitney test (presented in Appendix 4), which showed that none of the variables tested produced significant differences (at 5% significance level) between early and late respondents. This suggested that, although bias in the response may exist in the sample of questions tested, they were not a significant factor which could affect the conclusions about the variables being studied.

7.5.1.2 Exploring relationships between variables: chi-square, logistic regression and discriminant analysis

The basis of the analysis used to explore the relationship between the adoption of ICT generally and the adoption of CAT tools in particular was an expectation-based adoption model. This model searched for those translators most likely to adopt CAT tools depending on their adoption and degree of experience with other ICT. Logistic regression and discriminant analyses are two statistical methods often used for this kind of classification problem. Logistic regression builds a model to predict which category translators belong to based upon a set of predictors. Discriminant analysis takes the same approach but makes stronger assumptions about the predictor variables, specifically that the values of the variables follow a multivariate normal distribution with identical covariance matrices (Ye, 2003, p. 49). Based upon this, discriminant analysis is seldom appropriate since these assumptions are rarely met in practice. Logistic regression carries fewer assumptions than does discriminant analysis, particularly the ability to include categorical predictors, such as being the user of a particular software application. Logistic regression is used when it is important to predict whether a translator will adopt CAT tools or not based upon certain characteristics of the translator. It is particularly suitable where a binary (zero or one) or dichotomous dependent variable exists (e.g. in this case, translators who did not adopt CAT tools and translators who did adopt them).

Logistic regression estimates the coefficients of a probabilistic model, involving a set of independent variables in order to best predict the value of the dependent variable. A positive coefficient for an independent variable increases the predicted probability, while a negative value decreases the predicted probability of the outcome being in either of the two dependent variable categories (Hair et al., 1998, p. 130). In predicting the probability effects of multiple independent variables on a single dichotomous dependent variable, the model used is:

$$p(y=1) = \frac{1}{1+e^{-z}}$$

where:

$$z = \beta_0 + \beta_1 x_1 + \beta_2 x_2 + \ldots + \beta_n x_n$$

χi = an independent variable
$\beta 0$ = an intercept term
βi = the parameter for the independent variable χi
e = the quantity 2.71828 +, the base of natural logarithms
y = the dichotomous dependent variable, here CAT adoption
$p(y = 1)$ = the probability of a translator being classified as a CAT adopter.

The results obtained through the logistic regression model were then compared with those obtained through chi-square tests conducted individually between each of the nominal variables measuring ICT usage and the nominal dependent variable "CAT adopter: Yes/No."

Chi-square (χ^2) is a general test designed to evaluate whether the difference between observed frequencies and expected frequencies under a set of theoretical assumptions is statistically significant or simply random variation. This statistical test is most often applied to problems in which two nominal variables are cross-classified in a bivariate table (Frankfort-Nachmias and Nachmias, 1992, p. 464).

In order to investigate the relationships between the adoption of CAT tools and the level of knowledge of ICT, further chi-square tests were performed. The analysis included each of the nominal variables measuring the degree of familiarity and experience with the ICT available to translators and the nominal dependent variable "CAT adopter: Yes/No." To obtain a richer picture of the relationship of CAT tool adoption with the level of knowledge of the ICT for the activities in the translator's workflow, Internet-based ICT was also included in this part of the analysis. Since there were four possible values for the level of knowledge, these chi-square statistics should generally be compared with the chi-square distribution with three degrees of freedom.

In addition, chi-square was used to determine whether relationships existed between each of the characteristics of the sample and the group of CAT tool adopters, that is, between profile nominal variables (e.g. age range, gender, education) and the nominal dependent variable "CAT adopter: Yes/No."

7.5.1.3 Using factor analysis to measure the perceptions of CAT Tools

In order to achieve a better understanding of the structure of the data and to identify eventual underlying dimensions, factor analysis is a multivariate statistical method that analyses the interrelationships (correlations) among a number of items, and then determines the extent to which each variable is explained by each dimension, known as a factor (Hair et al., 1998, p. 90).

The main objective of factor analysis is to reduce the wide-ranging number of variables into more manageable groups of factors (Lehman, 1989). The technique assumes that there are only a few basic dimensions that underlie attributes of a certain construct to be measured and it then correlates the attributes to identify these basic dimensions (Churchill, 1999). Factor loadings produced from factor analysis are used to indicate the correlation between each attribute and each score, the higher the factor loading the more significant those attributes are in interpreting the factor matrix (Hair et al., 1998, p. 106).

To use factor analysis, a number of requirements need to be met. According to Sproull (1988), variables under study have at least to be of interval scale for factor analysis to be appropriately applied. In this study, the variables used measured the translators' perceptions of ICT through an ordinal scale. However, this does not preclude the use of factor analysis because an ordinal scale can be treated as an interval scale if one assumes that the distortion introduced by assigning numeric values to ordinal categories is not very substantial (Kim, 1975). Kim and Mueller (1978) indicated that many ordinal variables may be given numeric values without distorting the underlying properties, particularly, as in this case, when numeric values are shown on the questionnaire to guide respondents. Therefore, in this study, it was also assumed that the distortion effect, as a result of assigning numeric values to ordinal data, was not significant.

The Kaiser-Meyer-Olkin (KMO) Measure of Sampling Adequacy test and the Bartlett test of sphericity can be used to test whether it is appropriate to proceed with factor analysis. A small value on the KMO test indicates that the factor analysis may not be a good option. Kinnear and Gray (2000) suggest that the KMO value should be greater than 0.50 for the factor analysis to proceed. Kaiser (1974) suggests that a KMO measure in the 0.90s is considered to be "marvellous" sample adequacy for factor analysis purposes, in the 0.80s to be "meritorious," in the 0.70s is considered to be "middling," in the 0.60s is considered to be "mediocre," in the 0.50s is considered to be "miserable," and below 0.50s is considered to be "unacceptable."

The Bartlett test of sphericity and its significance level consider whether the variables are independent (i.e. form an identity matrix) and hence determine whether factor analysis is an appropriate technique to use. If the Bartlett test value is not significant (that is, its associated probability is greater than 0.05), then it is likely that the correlation matrix is an identity matrix (where the diagonal elements are 1 and the off diagonal elements are 0) and is therefore unsuitable for further analysis (Kinnear and Gray, 2000). What is required is that Bartlett's value for testing sphericity is large and the associated significance is small, that is, less than 0.05. When these criteria are present, the data are suitable for factor analysis.

7.5.1.4 Using ANOVA to compare CAT tool perceptions between adopters and non-adopters

A suitable way of comparing two or more variables for its statistical significance is using the analysis of variance (ANOVA). Its purpose is to determine whether a factor has a significant effect on the variable being used (CAT tool adoption, in our example) by comparing the means from several independent groups (Kvanli et al., 2002, p. 446). In the study, mean values of adopters and non-adopters of CAT tools were compared and their statistical significance assessed using ANOVA at the 0.05 significance level to examine how perceptions of CAT tools differed between these two groups.

7.5.2 A qualitative data analysis approach

A qualitative approach was deemed suitable to analyse the data of the second phase of the research to gain more insights into ICT adoption. More precisely, the qualitative data analysis framework suggested by Miles and Huberman (1994) for cross-case analysis was followed. One of the aims of this method is to draw conclusions from the multiple cases under study to increase generalisability and to reassure the researcher that the events and processes in one well-described setting are not wholly idiosyncratic. This allows developing more sophisticated descriptions and more powerful explanations from the data (idem, p. 172). In order to achieve this, the method involves three major activities: data reduction, data display, and conclusion drawing and verification (idem, p. 10), as represented in Figure 7.3.

A schematic view of how Miles and Huberman's method was applied to the study is presented in Figure 7.4 and an applied sample is included in the more detailed

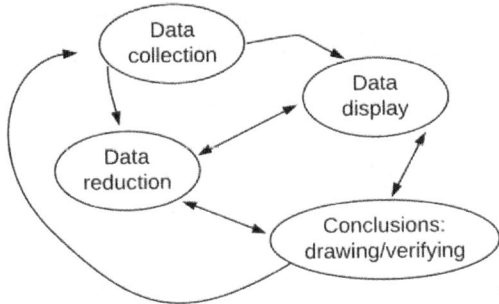

Figure 7.3 Components of Miles and Huberman data analysis method (1994, p. 12)

Qualitative Data Analysis Plan

Qualitative Data Analysis

DATA COLLECTION PERIOD

DATA REDUCTION

PRELIMINARY	DURING	POST
- Conceptual Frmwork - Cases - RQs - Collection approach	- Summaries - Coding - Testing themes - Make clusters/patterns & partitions	- Report production

DATA DISPLAYS

		DURING	POST
CROSS-CASE STRATEGY	Mixed (variable-case- oriented)	- Stacking comparable cases - Unordered META-MATRIX	- Partially ordered - Conceptually ordered (content-analytic)

CONCLUSION DRAWING/VERIFICATION

DURING	POST
- Inductive conclusions: noting patterns, explanations, regularities	- Verifying conclusions + writing "Final" ones

Figure 7.4 Qualitative data analysis plan

summary of the steps involved for analysing the particular issue of the relationship between continuous training and the use of specialist-purpose ICT in Appendix 5.

7.5.2.1 Data reduction

Data reduction helps to sharpen, sort, focus, discard, and organise the data in a way that allows for "final" conclusions to be drawn and verified. Data can be reduced and transformed through such means as selection, summary, paraphrasing, or through being subsumed in a larger pattern (idem, p. 10).

For each case (i.e. for each translator), the responses obtained for each construct (i.e. relative advantage, compatibility, voluntariness, image, ease of use, result demonstrability, visibility and trialability) were reduced by converting degrees of agreement to the numerical data and averaging them. As well as permitting the exploration of each respondent's views, the numeric conversion of the scale data facilitated the calculation of the relative frequency with which each issue was present, as well as the intensity with which they were expressed (as the scale values represented ordinal data).

7.5.2.2 Data display

Data display involves taking the reduced data and displaying it in an organised, compressed way so that conclusions can be more easily drawn (idem, p. 11). It allows analysts to understand what is happening and extrapolate enough information to discern systematic patterns and interrelationships. Miles and Huberman identified a number of valid types of display, namely, matrices, graphs, charts and networks, to assemble organised data into an immediately accessible form which facilitates its understanding for conclusion drawing or further analysis.

In this study, cross-case matrix data displays were deemed extremely helpful to show a comprehensive picture of the data collected for each case, i.e. within-case analysis of each translator (idem, p. 174), and to identify patterns across the constructs investigated to determine the factors affecting the adoption of CAT tools, i.e. variable-oriented analysis across the cases (idem, p. 175). The decision to adopt CAT tools was likely to vary from one translator to another, based on factors such as the respondents' perceptions of the adoption of CAT tools, the translators' personal characteristics and the differences in the translation businesses.

Each translator who participated represented a case (i.e. the rows of the displays), and the constructs used[11] represented the variables under study (i.e. the columns of the displays). A mixed strategy was followed to perform a cross-case analysis, combining case-oriented and variable-oriented approaches, which had been found to be a desirable way of analysing data from multiple cases (idem, p. 176). As noted by Ragin (1987), each approach has pluses and minuses: variable-oriented analysis is good for finding relationships among variables, but poor at handling the complexities of causation; and case-oriented analysis is good at finding specific patterns common to small sets of cases, but its findings often remain particularistic.

One strategy suggested by Miles and Huberman to follow a mixed approach for cross-case analysis is called by them "stacking comparable cases" (1994, p. 176). After series of cases have been entered for a set of variables, matrices are used to analyse each case in depth. When each case has been understood, it is "stacked" in a "meta-matrix," which presents a further condensed and comparable display. The study being used as an example followed a "stacking comparable cases" strategy to perform cross-case analysis.

---

[11] I.e. relative advantage, compatibility, voluntariness, image, ease of use, result demonstrability, visibility and trialability.

7.5.2.3 Conclusion drawing and verification

Conclusion drawing and verification is the activity in which the researcher begins to decide what things mean. He or she does this by noting regularities, patterns, explanations, possible configurations, casual flows and propositions (idem, p. 11).

After the data reduction and data display stages of the cross-case analysis, the last part of the analysis is focused on verifying and validating the quality of the preliminary conclusions drawn during the analysis of the displays. Some tactics can be employed to observe what was expressed by the data (e.g. *noting patterns, counting*), other tactics can be used to achieve a deeper understanding of what was being observed (e.g. *making contrasts/comparisons*), and other tactics help the researcher to see relationships more abstractly (e.g. *factoring, noting relations between variables*).

In this study, the identification of patterns across the cases for all the predictors was a tactic frequently used to see added evidence of recurring regularities, which is especially suitable when the number of cases and the data is large (idem, p. 246). In the matrix displays of the present study different shadings were used to identify the different degrees of importance of each of the predictors examined and to help to note patterns across the cases of each matrix.

Although a basic principle of qualitative research is to go beyond how much there is of something to find out what the essential qualities are, "a lot of counting goes in the background when judgements of qualities are being made" (idem, p. 253). For example, during the identification of patterns, the consistency or recurrence of facts is, in part, estimated by making counts, comparisons, and weights. As Miles and Huberman explain (idem, p. 253), numbers help to see rapidly what you have in a large batch of data; verifying a hunch or hypothesis; and keeping yourself analytically honest, protecting against bias. In the study, scales were converted to numerical scales which helped to use counts as a constant way of evaluating the real weight of patterns, and of ensuring that claims based on impressions were empirically grounded and not just personal intuitions.

Another useful tactic used in the cross-case analysis of this study was making contrasts and comparisons to test conclusions and assess the practical significance of the conclusions drawn. Comparisons and contrasts were especially useful to examine the two adoption settings (translators who had adopted, and translators who had not adopted CAT tools) against each of the predictor variables.

Verification of conclusions is the next step. After making and interpreting findings at different levels of inference, the representativeness and reliability of the findings need to be checked. Although the assessment of the "goodness" of qualitative research does not have a strong tradition as in quantitative research, Miles and Huberman (idem, p. 262) propose a number of tactics to help researchers at the operating level to test and confirm findings.

The representativeness of the data used to draw the conclusions is assessed by checking its quality (i.e. whether a finding is an instance of a more general phenomenon). Miles and Huberman identify three common pitfalls and their associated sources of error in making sure that the data collected is valid (idem, p. 264): "sampling nonrepresentative informants," "generalising from nonrepresentative events or activities," and "drawing inferences from nonrepresentative processes."

In the study, the sampling process involved a number of translators who had previously participated in the earlier fieldwork of the research. No selection process was undertaken to only address to a particular elite of informants (i.e. translators who might be responsive informants because they are experts in the field and thus, their opinions could be biased); on the contrary, the sample for this phase of the study included a relatively large number of individuals (151 translators) who were adopters or non-adopters of CAT tools. With regard to the other two common pitfalls, the use of a validated instrument to measure translators' perceptions (Moore and Benbasat, 1991), helped to have a number of constructs and processes from which conclusions were drawn, thus avoiding generalisations or drawing inferences from nonrepresentative events or processes.

Needs and perspectives of multilingual information professionals: findings of an empirical study

After introducing research methods and data analysis approaches to investigate the issues around the adoption of ICT in the professional context of multilingual information management, now the main findings of the study undertaken in previous empirical research (Fulford and Granell-Zafra, 2004; Fulford and Granell-Zafra, 2005; Granell-Zafra, 2006) are presented to provide an overview of the information and technology needs and perspectives in the context of multilingual information professionals.

8.1 Characteristics of MIPs

The research was based on a sample of 391 freelance translators. In comparison with other studies of translators, both the sample size generated for analysis and the response rate to this survey (35%) were encouraging. Out of this initial sample used with an exploratory purpose, a follow-up sub-sample of 53 translators (34 non-adopters and 19 adopters of CAT tools) was used to analyse in detail the adoption of specialised ICT and its impacts on their work, representing a response rate of 43% in this subsequent part of the study.

Some of the factors in the literature reviewed that were more importantly affecting the adoption of ICT in SMEs came from the characteristics of the CEO (c.f. Chapter 5, section 5.3.3.1). Therefore, in the context of freelance translators, the knowledge, training and capabilities of the translators are likely to affect the adoption of CAT tools. It was necessary to identify how the characteristics of freelance translators might be affecting this adoption.

Similarly, the organisational characteristics of a business have been found to affect the adoption of ICT in SMEs in the literature reviewed (c.f. Chapter 5.3.4). In the context of freelance translators, the characteristics of the translation business are likely to affect the adoption of CAT tools. It was necessary to identify how the characteristics of the freelance translation businesses might be affecting this adoption.

The following sections summarise the characteristics of both the translators' background information and their business, in addition to presenting an overall profile of ICT use of the translators in the sample of the study.

8.1.1 Age, experience and gender

Most of the respondents were aged between 30 and 59 (79%), with the highest proportion being in the 40-49 age group (30%). In addition, the vast majority of the respondents were aged 30 years or above (96%), thus showing that the sample was mostly made up of professionals. In fact, the age of the participants was in line with their professional level of experience and knowledge of the translation sector. The overriding majority of the respondents (80%) were quite established in the profession, as they had six or more years of experience, as opposed to 20% that had fewer than six years of experience and could be considered newcomers or still in the process of consolidating themselves within the profession. The median and mean values were 11 and 13 years, respectively.

The majority of the translators were female (63%), a fact that tallied with the information provided by Labour Force Survey in UK (ONS, 2003) with regard to self-employment, industries and gender for the years 1991 to 2003, which indicated that, although self-employment is more common among men than women, when looking at industries as a whole, men tend to be under-represented in the category that includes translation services.

8.1.2 Educational level and qualifications

The translators surveyed were highly educated, since the overriding majority (84%) of them had obtained a university degree (39% held a bachelor's and another 39% had attained a master's degree), and the remainder held postgraduate-level diplomas. Seventy-one per cent had a translation-related qualification (rather than simply a general qualification in languages) and some respondents had obtained more than one qualification in translation (i.e. a postgraduate degree in translation studies or a diploma in translation from a professional language or translation institute).

These findings also served to increase the reliability of the responses obtained, since a majority of the respondents were not only members of one or more UK language or translation professional bodies, but had followed formal training and held specific translation qualifications.

Further details about the educational background included any training undertaken to acquire ICT knowledge and skills, such as through taught courses or through private study. Thirty-eight per cent of the respondents had acquired their ICT knowledge and skills on a self-taught basis only. However, the vast majority of the respondents in the sample (85%) acquired their ICT knowledge and skills on a self-taught basis and through some other type of ICT training, namely, professional training courses (37%), workshops run by professional institutes (18%), university or college courses (15%), or individual IT modules as part of a university degree (8%), or in-house training (8%).

Only 18% of the respondents held a formal ICT-related qualification, the majority of which were professional ICT qualifications/certificates, such as ECDL.[1] A small number (8%) had university degrees in computing-related areas, and some (23%) had undertaken ICT-related assessed modules as part of other degree programmes.

8.1.3 Translation as business

The translation work carried out by each respondent was measured in terms of the number of words translated per week (volume of translation) and the number of

---

[1] ECDL is the international standard in digital skills certification. Further details can be found at *http://www.ecdl.com*

hours dedicated to translation work per week (hours worked). The average workload was 6,000 words per week, and the most usual volume (i.e. the mode) was 10,000 words per week. The usual dedication was full-time (i.e. 40 hours/week), although the median value decreased to 25 hours per week, thus indicating that a large number of translators were part time workers.

In terms of the subject area where translation services were provided, business/commerce (79%) was the most common practice, and more than half of the respondents were working in the areas of technical (55%) and legal (53%) translation. Other areas included financial (40%), scientific (28%), literary (20%), medical (8%) and academic translation (3%).

Language combinations of the multilingual communication services provided by respondents included a range of possibilities, although the main trends were that almost half of the respondents (47%) were translating from an EU official language into English, while 13% were translating in the reverse direction, and 14% were translating these languages in both directions. The most common EU source languages that respondents were translating into English were German, French and Spanish.

8.2 ICT adoption

Overall, the ICT used by the translators in the sample was mostly supporting three types of activity: communication activities (47.8%), document production activities (47.1%), and information search and retrieval activities (46.8%). After this group of three activities, ICT was used by less than a third of the translators to support business management activities (29.7%), marketing and work procurement activities (22.2%), and translation creation activities (9.8%). These figures show that the breadth of use of general-purpose software and online facilities by translators (i.e. ICT used within the first three types of activity) was higher than the breadth of use of ICT to support other activities for which more specialised software is available (e.g. accounting software, translation marketplaces or CAT tools). Therefore, there was a progression from a greater breadth of ICT usage for activities which required more general-purpose ICT to a smaller breadth of ICT usage for activities which required a more specialist type of ICT.

At the same time, a similar progression can be observed within the range of ICT to support each of the activities, especially with regard to those closer to the nature of the translation work (i.e. information search and retrieval, and translation creation). ICT was used by a greater proportion of the respondents in those types of activity with a more general use (e.g. online search engines or online dictionaries) than specialist-purpose applications (e.g. terminology management systems). Even within the translation creation activities category, more specific types of applications, like localisation software (used by 2.3% of the respondents), were used by a much smaller proportion of the translators than CAT tools (28.3%).

Table 8.1 presents a summary of the findings on ICT usage and the activities supported, ranked in descending order of ICT usage.

Table 8.1 ICT usage

Software application / online facility	Frequency (F)	Total cases (N)	ICT usage
Communication			
Email	350	376	93.1%
Discussion mailing lists	118	320	36.9%
FTP (File Transfer Protocol)	102	320	31.9%
Online discussion groups	92	315	29.2%
Total activity usage	*662*	*1331*	*47.8%*
Document production			
Word processing software	386	391	98.7%
Graphical / presentation software	82	325	25.2%
Desktop Publishing software	57	328	17.4%
Total activity usage	*525*	*1044*	*47.1%*
Information search and retrieval			
Online search engines	316	372	84.9%
Online dictionaries / glossaries	292	372	78.5%
Online multilingual terminology databanks	202	344	58.7%
Text corpora / document archives	172	339	50.7%
Online encyclopaedia	126	331	38.1%
Academic journals	98	324	30.2%
Electronic databases	94	318	29.6%
Electronic libraries	85	320	26.6%
Terminology management systems	77	322	23.9%
Total activity usage	*1462*	*3042*	*46.8%*
Business management			
Spreadsheet software	291	370	78.6%
Database software	84	335	25.1%
Accounting / bookkeeping software	42	317	13.2%
Project and workflow management software	6	309	1.9%
Total activity usage	*423*	*1331*	*29.7%*
Marketing and work procurement			
Online translation marketplaces	103	316	32.6%
Own web site offering translation services	82	391	21.0%
Web publishing software	41	316	13.0%
Total activity usage	*226*	*1023*	*22.2%*
Translation creation			
CAT tools	94	332	28.3%
Machine translation systems	16	313	5.1%
Online machine translation services	11	308	3.6%
Localisation software	7	301	2.3%
Total activity usage	*128*	*1254*	*9.8%*

8.2.1 Familiarity and experience with ICT

In addition to indicating the respondents' actual usage of ICT, they were asked to indicate their degree of familiarity and experience with each of the types of tool/online facility discussed for each activity in the translator's workflow. This was measured by a scale of familiarity and experience (namely, "Not familiar," "Familiar, but with no working knowledge," "Familiar, with some working knowledge," and "Familiar, with extensive working knowledge").

8.2.1.1 Document production activities

Word processing software was not only the most widespread type of ICT used by the respondents, but also the one they had most experience with (89% having extensive experience and 11% having some experience with it). With regard to the other two types of software supporting document production activities (i.e. graphical or presentation software and desktop publishing software), which were used by less than a quarter of the translators, around 42% of the respondents were not familiar with them, and around 22% were familiar with them but had no experience using them.

8.2.1.2 Information search and retrieval activities

Online search engines were in widespread use (85%) and just a few translators were not familiar with them at all (4%) or knew about them but had no experience using them (3%). Among those who had experience using them, most (72%) had extensive experience with this online facility. Most of the translators in the sample had some (43%) or extensive (44%) experience with online dictionaries and/or glossaries, which were being used by 79% of the respondents. Text corpora/document archives (used by around half of the translators) followed, with translators having some (41%) or extensive (30%) experience, although they were unknown to 14% of the respondents, and 15% were aware of them but had no experience with them.

Multilingual terminology databanks, which were used more than text corpora/document archives (59%), were slightly less known to the translators (19% had no familiarity with this online facility), and 13% knew of, but had no experience with them. Degrees of experience with multilingual terminology databanks were similar to experience with text corpora/document archives: 39% had some experience and 30% had extensive experience with them.

Other online reference resources that were used by around a third of the respondents were unfamiliar to about a third of them: 26% were unaware of online encyclopaedias, 28% of online academic journals, 35% of electronic databases and 41% of electronic libraries. Also, among those who were familiar with these online facilities, 19% had no experience with online encyclopaedias, 22% with online academic journals, 21% with electronic databases and 20% with electronic libraries. As shown in detail in Table 5.13, around a third of those who did have experience with these facilities had some experience with them, and around 17% had extensive experience with them.

With regard to terminology management systems, the least used ICT (by 24%) among those supporting information and search retrieval, half of the translators were not familiar with these tools at all, and 21% were aware of but had no experience with them. Among the translators with experience, 19% had some and 10% had extensive experience.

Overall, the computer tools supporting the information search and retrieval activities with which translators were more experienced were online search engines and online dictionaries and/or glossaries (technologies which were in widespread use), followed by ICT such as text corpora/document archives, multilingual terminology databanks (which were used by about half of the respondents). After these ICT, translators were less experienced with a group of technologies used by around a third of the respondents (namely, online encyclopaedias, online academic journals, electronic databases and electronic libraries), leaving terminology management tools as the least used and least familiar type of ICT supporting information search and retrieval activities. This reflected an overall better knowledge and usage of more general ICT, as opposed to lower levels of awareness and experience with ICT particularly designed to support translators' activity (with terminology management systems as the most clear example).

8.2.1.3 Business management activities

Spreadsheet packages, used by 79% of the respondents, were not familiar to 6% of them, and another 8% had no experience in working with them. The majority of the respondents (57%) had some or extensive (29%) experience with spreadsheet packages.

On the other hand, the rest of the ICT supporting business activities were more unfamiliar to the translators: database packages were not familiar to 34% of the respondents, dedicated accounting/financial management packages to 65% of them, and the vast majority (87%) were unaware of project management software. These types of software were familiar to less than a third of the respondents who had no experience with them: 28% (databases), 18% (accounting software) and 8% (project management), respectively. Translators had more experience with databases (31% with some and 8% with extensive experience) than with the other accounting/financial packages (12% with some, and 5% with extensive experience), and just 4% had some experience with project management software.

8.2.1.4 Translation creation activities

Twenty-eight per cent of the respondents were using CAT tools, around 18% had some experience, and 18% had extensive experience with them. Forty per cent of the translators were not familiar at all with this type of technology, while 25% were aware of them but had no experience in using them. The rest of ICT supporting "translation creation activities" (i.e. machine translation systems, online machine translation services and localisation software) were used by a minority (between 2% and 5% of the respondents), and the majority of the respondents (between 64% and 91%) were not familiar with these technologies. There was very little evidence of experience (and to a small extent) with them, 11% had experience with machine translation systems, 7% had experience with online machine translation services and 4% had experience with localisation software.

8.2.1.5 Communication activities

Electronic mail was not only in widespread use, but the vast majority of the respondents (89%) also had plenty of experience with it. Around two thirds of the respondents did not make use of online mailing lists and discussion groups, and around 36% had no familiarity with them. Around 20% of the translators were familiar with these online facilities but had no experience with them, and about 40% did have some degree of experience with electronic mailing lists or online discussion groups for translators. Thirty-two per cent of the translators used FTP software, but 34% of them were not familiar with this type of software; 18% knew about it, but had no experience, and around half of the translators had experience with FTP applications.

8.2.1.6 Marketing and work procurement activities

Sixty-six per cent of the respondents were not familiar at all with web publishing software, used by only 13% of the respondents. Another 18% were familiar but had no experience with this type of application, and just 16% had some (12%) or extensive (4%) experience with it. Online translation marketplaces were being used by around a third of the translators; however, almost half of them (43%) were not familiar at all with these online facilities for marketing and/or work procurement. Eighteen per cent of the translators were familiar with them but had no experience, and almost 40% had some (22%) or extensive (17%) experience with them.

Table 8.2 summarises the findings on ICT familiarity and experience for the activities supported, showing the percentage of responses for each degree of familiarity and experience, and the mean of the value obtained from the scale measuring this degree. Activities and ICT within each activity are ranked in descending order of experience.

8.2.2 Relationship between ICT usage and familiarity and experience with ICT

Once the findings of ICT usage and the degree of familiarity and experience of the respondents were observed in relation to the activities that are part of the translator's workflow, a more detailed analysis was undertaken using the statistical technique of cluster analysis. This technique facilitated the allocation of individual translators to one of several clusters (or groups) in which their members tended to share a number of characteristics in common with other members of the cluster (or were considered in statistical terms to be closely aligned to that cluster).

Based on this analysis, two broad groups of activities were identified (Figure 8.1):

- On one side, communication (A1), document production (A2), and information search and retrieval (A3), which accounted for higher ICT usage and higher experience with ICT (going towards extensive usage and experience);

Table 8.2 Familiarity and experience with ICT

Type of software application / online facility	Not familiar	Familiar but no experience	Familiar with some experience	Familiar with extensive experience	Total sresponses	Scale mean
Communication						
Email	1.7	0	9.6	88.7	354	3.85
FTP (File Transfer Protocol)	33.5	17.7	27.8	20.9	316	2.36
Discussion mailing lists	38.1	18.3	22.1	21.5	312	2.27
Online discussion groups	35.2	24.8	22.3	17.7	310	2.23
Activity average mean						2.68
Information search and retrieval						
Online search engines	3.7	3.1	21.6	71.6	356	3.61
Online dictionaries / glossaries	4.2	9.1	43.2	43.5	361	3.26
Text corpora / document archives	14.2	14.5	40.9	30.3	330	2.87
Online terminology databanks	18.6	12.9	38.9	29.6	334	2.80
Online encyclopedias	25.8	18.7	35.3	20.2	326	2.50
Academic journals	27.9	22.2	32.7	17.1	315	2.39
Electronic databases	35.4	21.1	26.6	16.9	308	2.25
Electronic libraries	40.7	19.9	22.1	17.3	312	2.16
Terminology management systems	50	20.8	18.9	10.2	322	1.89
Activity average mean						2.64
Document production						
Word processing software	0.5	0	10.7	88.8	366	3.88
Graphical / presentation software	41.6	23	28.3	7.1	322	2.01
Desktop Publishing software	43.7	21.5	27.7	7.1	325	1.98
Activity average mean						2.62

Business management					
Spreadsheet software	6.5	7.9	57	28.7	356	3.08
Database software	33.5	27.5	30.8	8.2	331	2.14
Accounting / bookkeeping software	64.8	17.8	12.4	5.1	315	1.58
Project management software	87.1	8.4	4.2	0.3	309	1.18
Activity average mean						*1.99*
Marketing and work procurement					
Online translation marketplaces	43.6	17.6	21.8	17	312	2.12
Web publishing software	65.9	18.2	11.8	4.1	314	1.54
Activity average mean						*1.83*
Translation creation					
CAT tools	39.2	24.9	17.6	18.2	329	2.15
Online machine translation services	64.3	24.7	8.8	2.3	308	1.49
Machine translation systems	75	17.6	5.1	2.2	312	1.35
Localisation software	90.7	5.7	2.7	1	300	1.14
Activity average mean						*1.53*

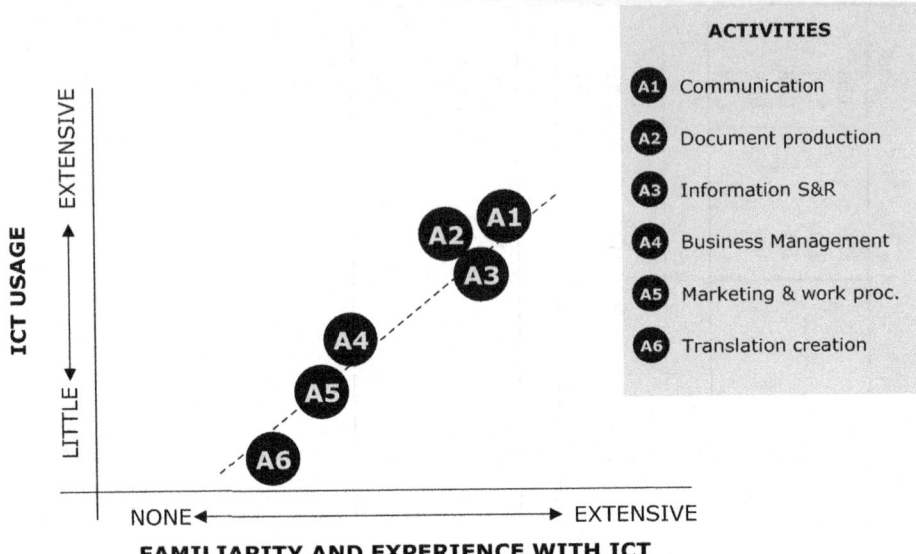

Figure 8.1 ICT usage and familiarity and experience with ICT

- And, on the other side, business management (A4), marketing and work procurement (A5) and translation creation (A6), which accounted for little ICT usage and little familiarity and experience.

Also, it should be noticed that A5, and especially A3 and A6 were positioned slightly lower than an imaginary diagonal line that would show an exact correlation between usage and experience. This was reflecting those activities with software that were specifically developed for translators (e.g. CAT tools, terminology management systems and translation marketplaces) with which the translators had more experience than should match with their actual usage. This implied that some translators might not actually be using these types of translation specialised software although they were experienced with them. Therefore, this might reflect a sense of "craftsmanship" about translation, and a reluctance to use technologies that may automate certain tasks.

8.3 CAT tool adoption

According to the findings of this study, only 28% of the respondents in a sample of 391 freelance translators had adopted CAT tools, which indicated a key issue of this research: the level of adoption of CAT tools among freelancers was not very high. This finding by this study revealed that, contrary to what has been reported in the literature (see for example Somers, 2003a, p. 6), the adoption of CAT tools did not seem to be widespread among freelancers in the UK.

The findings further revealed that 39% of the respondents were not familiar at all with CAT tools, which revealed a significant proportion of translators who were not familiar with the capabilities and benefits that they could obtain from their use of these tools. Twenty-five per cent of the respondents were familiar with CAT tools but had no experience with them, maybe due to a lack of opportunities to try them out, or other deterrents from adopting them. Finally, 36% of the freelancers had some or extensive experience with them, a slightly larger proportion of the 28% of the translators who had adopted a CAT tool, meaning that some of the translators in the sample were not currently using CAT tools in spite of having some degree of experience with them. This might reflect a voluntary non-adoption of CAT tools which might be due to a number of reasons (such as the incompatibility with the work translators undertake, or the high difficulty of learning to use these tools perceived by non-adopters).

From the findings about the levels of familiarity and experience with CAT tools, it could be concluded that the non-adoption of CAT tools was largely due to a lack of awareness of the capabilities of CAT tools for almost half of the translators in the sample, but there was also a significant proportion of freelancers whose levels of awareness of CAT tools were higher and yet they had not still adopted them. The analysis of the determinants for the non-adoption of CAT tools is presented later in this chapter.

From the findings, there was evidence of scepticism among freelancers about the value of CAT tools, and a lack of confidence in the benefits that might be gained from using them. While the reasons for this were no doubt many and varied (e.g. the perceived suitability of CAT tools to some freelance translators' type of work, and a fear of the difficulty of learning to use CAT tools among non-adopters of these technologies), institutions training translators, professional bodies of translators, and translation software developers had a role to play in raising awareness about CAT tools. If the awareness levels of CAT tools among the freelance translation community increased, it would be easier for freelancers to make an informed decision about the adoption of ICT supporting the core activity of a translation business (i.e. translation creation), such as CAT tools.

One possible way to increase uptake might be for CAT tool developers and/or distributors to heed the advice given by Martin Kay in his report on computers and translation (Kay, 1980; reproduced 1997). In his report, Kay advocated a gradual, step-by-step approach to adding tools into the translator's workstation, thereby slowly increasing the software support introduced into the translator's workflow. He stressed the importance of accepting an individual tool only once there is a reasonable degree of confidence about its capabilities, and its reliability in performing the tasks it is designed to support. By contrast, the tendency with much software development today – and CAT tool development is no exception – is to produce integrated packages or "bundles" of several tools. Typically, the entire package must be purchased in one go, giving little opportunity for the sort of incremental adoption that Kay advocated.

8.3.1 CAT tool adoption, other ICT adoption and freelance translators' activities

The limited uptake of CAT tools reported in the previous sections is related to the main activity in the translator's workflow, namely the production of translations. However, freelance translators today do not only have to produce translations, but there are also other activities in their workflow for which ICT support is available. As recognised by

Austermühl, the ICT that translators use must encompass both the notion of "translation as a business" as well as translation "as a linguistic and cultural process" (Austermühl, 2001, p. 11). The uptake of ICT available for other activities was also investigated in relation with the adoption of CAT tools.

The findings of the survey indicated that there was widespread adoption of general-purpose software applications to support the activities involved in the freelance translator's workflow, with different degrees of intensity. There was, however, only limited uptake of other specialised translation-oriented software applications, such as terminology management systems, machine translation, localisation software and, as discussed above, CAT tools. Likewise, there was only limited adoption of specialised software to support such business functions as financial management and accounting.

Overall, ICT uptake was more widespread among activities which were not exclusive to translation (i.e. communication, document production and information search and retrieval) than it was among freelance translator-specific activities (i.e. translation marketing and work procurement, translation creation and business management). A more detailed observation of ICT uptake within translators' activities also showed that the more specific a tool was to translation (e.g. terminology management systems), the lower the level of usage that could be observed. This broader range of ICT use for activities that were not specific to the translation context, and the limited uptake of ICT specially designed for translators revealed that, although translators are using technologies to support their work, their use is geared towards undertaking general tasks which require ICT (e.g. using email to communicate with their clients, or using a word processor or spreadsheet software for producing electronic documents), rather than using ICT specifically designed for translators.

Unlike some evidence that was found of the voluntary non-adoption of CAT tools by a small proportion of the translators in the sample, it seemed likely that non-adoption of ICT in general was more a function of translators' lack of awareness of, and familiarity with, the types of software than an active rejection decision based on a thorough knowledge of the tools and their functionality. These unawareness levels seemed to foster scepticism about adopting new ICT and made translators follow a cautious approach towards incorporating technologies into their work. In addition, translators' familiarity with ICT showed that overall non-familiarity levels were increasing progressively within each group of activities as the type of ICT was becoming more specialised to the translation context. This could be explained by the fact that more general purpose ICT is part of the basic ICT knowledge that translators (and other users of ICT) have, and that existing efforts towards informing translators about ICT available to them should place more emphasis on the suitability and benefits of using ICT tailored to their needs.

8.3.2 Predicting CAT tools adoption in relation to ICT through a logistic regression model and chi-square tests

With regard to the relationship between the adoption of CAT tools and the adoption of other ICT, the findings of the survey showed that CAT tool adopters were using a broader range of ICT and had more experience with general ICT than those who had not adopted CAT tools. This idea was reinforced by the results obtained from the logistic regression model used to analyse the relationship between the adoption of the range of ICT and the adoption of CAT tools. Stand-alone terminology management systems, both in terms of uptake and experience with them, were the type of ICT which showed stronger links with the adoption of CAT tools. This made sense as most CAT tools include terminology management functions bundled in them, so translators who are familiar with these translation-specific tools are more likely to be familiar with CAT tools as well.

The logistic regression model used to analyse the relationship between the adoption of ICT and the adoption of CAT tools revealed that, with an overall accuracy of 89.2%, adoption of CAT tools was mostly determined by the usage of, and experience with, terminology management tools, usage of and experience with, graphics applications, and usage of spreadsheets, with the usage of terminology management tools being the most influential variable of these.[2] Looking at the CAT tool usage variable, of the 391 translators, 94 (24%) could be classified as CAT tools users, while 238 (61%) could be classified as non-users, and the remaining 59 (15%) constituted missing values. Multiple logistic regression analysis was undertaken using this dichotomous CAT user variable (ignoring the missing values) as the dependent variable, and the variables on the usage of and degree of experience with the rest of the software applications as the independent variables. A total of 279 of the 391 cases were used to estimate the model. One hundred and twelve cases were not included because they contained missing data for one or more of the variables.

The first step of the logistic regression analysis included the following variables in the model: word processing usage and experience, spreadsheet usage and experience, database usage and experience, computer-based accounting usage and experience, desktop publishing usage and experience, web publishing usage and experience, graphics usage and experience, information retrieval tool usage and experience, groupware usage and experience, project and workflow management usage and experience, terminology management usage and experience, machine translation usage and experience, and localisation usage and experience. Once the variables were entered, backward elimination was used to remove those which were not significantly related to CAT tool adoption. Table 8.3 presents the statistics of the logistic regression prediction model.

Classification Table for CAT tool usage[a]

Observed			Predicted		
			CAT user		Percentage Correct
			No	Yes	
CAT user	No		197	15	92.9
	Yes		15	52	77.6
Overall Percentage					89.2

[a] - The outvalue is .300

[2] Non-adopters were correctly classified in 92.9% of the cases, while adopters were correctly assigned in 77.6% of the cases. This seemed to offer a very good prediction of the adoption of CAT tools based upon the use of and experience with spreadsheet software, terminology management tools, graphics software, and localisation software (nine out of ten CAT tool adopters would be using these ICT), within an overall high percentage of accuracy (89.2%) of the prediction model developed.

Table 8.3 **CAT tool vs. ICT adoption prediction (logistic regression model)**[3]

		Variables in the Equation					
		B	S.E.	Wald	df	Sig.	Exp(B)
Step[a] 20	Spreadsheet usage	1.935	.768	6.345	1	.012	.966
	Graphics experience	-1.028	.356	8.313	1	.004	-.298
	Graphics usage	1.883	.714	6.965	1	.008	2.717
	Terminology Mgment experience	.724	.316	5.266	1	.022	1.935
	Terminology Mgment usage	2.886	.702	16.876	1	.000	17.919
	Localisation experience	1.235	.699	3.123	1	.077	3.438
	Constant	-5.108	1.182	18.666	1	.000	.006

[a] - Variable(s) entered on step 1: Word processing usage and experience, Spreadsheet usage and experience, Database usage and experience, Computer-based accounting usage and experience, Desktop publishing usage and experience, Web publishing usage and experience, Graphics us. experience, Information lettrieval tools, Groupware usage and experience, Project and workflow management usage and experience, Terminology management usage and experience, Machine translation usage and experience, Localisation usage and experience.

These results were then compared with those obtained through individual chi-square tests conducted for each of the ICT in turn (c.f. Table 8.4), and both analyses presented similar results, thus supporting the prediction made by the logistic regression model. Terminology management systems, both in terms of adoption and also in terms of experience with them, were the type of ICT which showed strongest links with the adoption of CAT tools.

The relationship between the degree of experience with the ICT for each activity in the translator's workflow and the adoption of CAT tools was also investigated through the use of chi-square tests (c.f. Table 8.5). Overall, experience with ICT for communication, information search and retrieval, business management, marketing and work procurement, and translation creation activity were found to be significantly related to the adoption of CAT tools (p values ≤ 0.05 in bold). Only experience with ICT for the document production activity did not present a strong link with the

[3]The column headed "B" contains the logistic regression coefficients. The second column (S.E.) contains the standard errors for the "B" coefficients. The Wald statistic was used to test whether the predictor variables were significantly related to the outcome measure (i.e. the adoption of CAT tools) adjusting for the other variables in the model. Generally the Wald statistic has a chi-squared distribution with one degree of freedom. The column Exp (B) presents the "B" coefficient raised to the exponential power, and these coefficients can be interpreted in terms of an odds shift in the outcome.

Table 8.4 CAT users and use of other ICT

Type of ICT	Chi-Square	Significance
Communication activity		
FTP (File Transfer Protocol)	24.171	0.000
Information search and retrieval activity		
Terminology management systems	167.665	0.000
Document production activity		
Word processing software	2.005	0.157
Graphical / presentation software	18.539	0.000
Desktop Publishing software	14.896	0.000
Business management activity		
Spreadsheet software	21.389	0.000
Database software	2.644	0.104
Accounting / bookkeeping software	1.582	0.208
Project management software	0.222	0.638
Marketing and work procurement activity		
Web publishing software	8.813	0.003
Translation creation activity		
Machine translation systems	11.846	0.001
Localisation software	21.984	0.000

adoption of CAT tools. This might be due to the fact that almost all translators were using word processing software for document production. In particular, those translators not using CAT tools (i.e. undertaking their translations in a more "traditional" way) would also be mostly using word processing software.

These findings were further considered through the comparison of the mean values of adopters and non-adopters of CAT tools in relation to their degree of experience with ICT for each activity in the translator's workflow. It seemed that CAT tool adopters had more experience with ICT for other activities in their workflow. This implied that, generally, those translators who had more experience with, and were more confident with, general purpose ICT were more likely to adopt CAT tools. This conclusion was also supported by the findings of the specific predictors of CAT tool adoption (see logistic regression analysis).

The main differences between the groups of adopters and non-adopters of CAT tools were observed in the experience with ICT that showed a more significant relationship with CAT tool adoption according to the chi-square tests conducted; for example, with terminology management systems, online translation marketplaces or online terminology databanks. Again, the relationship of CAT tool adoption with specialist purpose ICT reinforced the idea that freelance translators were more likely to embrace CAT tools once they had become familiar with general purpose ICT first, and then with other specialised ICT.

Table 8.5 CAT users and familiarity with other ICT

Type of ICT	Chi-Square	Significance	Scale mean
Communication activity			
Email	5.291	0.071	3.85
FTP (File Transfer Protocol)	28.762	0.000	2.36
Discussion mailing lists	14.796	0.002	2.27
Online discussion groups	15.099	0.002	2.23
Activity average mean			*2.68*
Information search and retrieval activity			
Online search engines	17.707	0.001	3.61
Online dictionaries / glossaries	17.810	0.000	3.26
Text corpora / document archives	2.902	0.407	2.87
Online terminology databanks	26.573	0.000	2.80
Online encyclopedias	9.455	0.024	2.50
Academic journals	1.764	0.623	2.39
Electronic databases	8.859	0.031	2.25
Electronic libraries	6.952	0.073	2.16
Terminology management systems	126.313	0.000	1.89
Activity average mean			*2.64*
Document production activity			
Word processing software	1.860	0.395	3.88
Graphical / presentation software	8.440	0.038	2.01
Desktop publishing software	7.268	0.064	1.98
Activity average mean			*2.62*
Business management activity			
Spreadsheet software	10.718	0.013	3.08
Database software	13.923	0.003	2.14
Accounting / bookkeeping software	5.474	0.140	1.58
Project management software	14.625	0.002	1.18
Activity average mean			*1.99*
Marketing and work procurement activity			
Online translation marketplaces	22.999	0.000	2.12
Web publishing software	9.724	0.021	1.54
Activity average mean			*1.83*
Translation creation activity			
Online machine translation services	7.477	0.058	1.49
Machine translation systems	15.114	0.002	1.35
Localisation software	36.129	0.000	1.14
Activity average mean			*1.53*

8.4 The characteristics of freelance translators adopting CAT tools

Literature in the area of IS adoption by SMEs has shown CEO involvement and enthusiasm towards technology to be one of the most important determinants of the decision to adopt ICT (see for example Cragg and King, 1993), and of the success in the use of the systems (DeLone, 1988). In freelance translation businesses, the manager is also the end-user of the technology, and therefore freelance translators do not only decide on the adoption of CAT tools but also have to use these tools.

After conducting chi-square and logistic regression analyses to examine the characteristics of the translators who were adopters of CAT tools, the findings revealed that CAT adopters tend to be young and tend to have obtained a university degree in translation studies. On the other hand, there was no association found with gender, their educational level, the length of experience as a translator or their IT qualifications.

Therefore, the findings of the study showed that there was a statistical association between the adoption of CAT tools and some characteristics of the freelance translators, including the fact that adopters tended to be young translators (chi-square statistic: value 9.951 and significance of 0.041), holding a university degree in translation studies (either undergraduate or postgraduate) (chi-square statistic: value 14.659 and significance of 0.012), whereas no statistical association was found between the educational level (chi-square statistic: value 1.293 and significance of 0.731), ICT knowledge acquisition (most of the translators had acquired their ICT skills through private study) (chi-square significance of 0.729) or the years of experience (chi-square statistic: value 9.716 and significance of 0.084) and CAT tool adoption.

These findings seemed to indicate that there was a likely connection between translators who have relatively recently undertaken translation studies at a higher education institution and the adoption of CAT tools. Since there was no link found between ICT knowledge acquisition and CAT tool adoption, the findings relating to the characteristics of freelance translators adopting CAT tools pointed towards formal training as a more significant determinant of CAT tool adoption, as opposed to self-taught learners. This can be understood as an indicator of the role that higher education institutions may have in providing translators with the appropriate knowledge to make them aware of the ICT available to them and how to use it.

8.5 The characteristics of the freelance translation businesses adopting CAT tools

As with other similar SMEs that need to use ICT, freelance translators are advised by previous research and professional associations to plan and define their requirements for ICT (see, for example, Proudlock et al., 1999). The strategies to achieve different levels of sophistication in the adoption of ICT depend on the characteristics of the translation business that affect the adoption of CAT tools. From the literature reviewed, a number of factors affecting the success of ICT in SMEs have been discussed in Chapter 6, section 6.3.2. Characteristics such as the attitudes of the users of the ICT or the CEO support and attitude towards ICT adoption (the user of CAT tools and the CEO of the freelance translation business being the same person) were investigated in our work and the findings provided evidence of some characteristics of the freelance translation businesses that are likely to be associated with the adoption of CAT tools. Contrary to what had been claimed by authors like Heyn (1998, p.123), who stated that "CAT tools are now used in almost every type of translation work: political, administrative, technical, advertising and biographical," the findings of this study only provided evidence of an association between freelance translation businesses and CAT tool adoption in which translators undertake technical translations. The type of translation jobs in this subject area usually includes documents with repetitive structures, and where frequent updates or revisions are required to be performed while maintaining terminological consistency. These document characteristics are among the ones that have been suggested as most suitable for CAT tools use (Bowker, 2002, p. 112). It could be said that although there was significant evidence of CAT tool adoption among freelance translation businesses run by translators working in the technical subject area, the adoption of CAT tools was not common among translation businesses with translators working in the subject areas claimed by Heyn.

Chi-square and logistic regression analyses to examine the characteristics of the translation businesses that had adopted CAT tools estimated that the length of experience, the volume of work undertaken per week, the number of hours of work per week, the undertaking of technical translation, and the language combinations of English and other EU languages were determinants of whether CAT tools were adopted or not (Table 8.6). Among these characteristics, it was important to highlight that the most significant variable was found to be undertaking technical translation, which could not be tested through chi-square tests. With the exception of the subject area variable, the findings obtained for the two methods were broadly the same, with the volume of work undertaken per week, the number of hours worked per week, and the translation between English and other EU languages as significant characteristics of the CAT tool adopters. On the other hand, the length of experience of the translators, which was found to be significant by the logistic regression analysis, was not found to be significant by the chi-square tests.

Classification Table[a]

		Predicted		
		Software apps USAGE: CAT		Percentage Correct
Observed		No	Yes	
Software apps	No	129	58	69.0
USAGE: CAT	Yes	18	73	80.2
Overall Percentage				72.7

[a] - The cut value is .300

Table 8.6 CAT tool adopters' characteristics (logistic regression model)[4]

		Variable in the equation					
		B	S.E.	Wald	df	Sig	Exp(B)
Step[a] 11	Length of experience	-1.005	.282	12.655	1	.000	.366
	Volume of workper week	.942	.266	12.526	1	.000	2.565
	Hours per week	.830	.257	10.440	1	.001	2.292
	Languages translated			9.921	7	.193	
	Languages translated(1)	3.026	1.194	6.426	1	.011	20.615
	Languages translated(2)	2.391	1.209	3.911	1	.048	10.929
	Languages translated(3)	2.475	1.371	3.260	1	.071	11.876
	Languages translated(4)	2.469	1.154	4.579	1	.032	11.816
	Languages translated(5)	1.845	1.244	2.201	1	.138	6.328
	Languages translated(6)	1.343	1.453	.855	1	.355	3.832
	Languages translated(7)	-5.357	17.474	.094	1	.759	.005
	Subjectarea: technical	1.381	.349	15.675	1	.000	3.980
	Constant	-5.434	1.297	17.541	1	.000	.004

[a] - Variable(s) entered on step 1: age, gender, educational background, translation qualifications, length of experience, volume of work per week, hours working per week, language pairs translated, subject areas translated, IT qualifications.

[4] The independent variables with a strong positive influence on the model were found to be length of experience, volume of work undertaken per week, hours of work undertaken per week, the subject area of technical translation, and the language combinations of English and other EU languages. The most influential variable was subject area of technical translation with a Wald score of 15.675 and a significance of 0.000. Other variables with a high Wald score and a significance of 0.000 were length of experience (Wald = 12.655) and volume of work undertaken per week (Wald = 12.526). The variables that were not statistically significant at the 5% confidence level were left out of the final model results by the logistic regression analysis. The results obtained through the logistic regression model were broadly similar to those obtained through the chi-square tests. The overall accuracy of the CAT tool adoption model was 72.7%. Non-adopters were correctly classified in 69% of the cases, while adopters were correctly assigned in 80.2% of the cases. This seemed to offer a good prediction of the adopters' business characteristics (four out of five CAT tool adopters would fit within these characteristics), within an overall high percentage of accuracy (72.7%) of the prediction model developed.

8.6 Perceptions of ICT and perceptions of CAT Tools

Views on CAT tools and on other ICT were investigated to try to understand the factors that affected CAT tool adoption. The existing scepticism about CAT tools reported in the literature (see, for example, Heyn, 1998; Hutchins, 1999) and in informal discussions was reflected in this study in the perceptions that translators had of these tools.

In trying to understand further the differences between the perceptions of CAT tools among adopters and non-adopters, the findings revealed that adopters' perceptions were overall much more positive than non-adopters. In particular, two issues emerged as important from the analysis of the perceptions of adopters and non-adopters: first, the former thought that CAT tools increase translators' effectiveness, while the latter thought the opposite; and second, adopters did not consider their use of CAT tools a failure at all. These issues seemed to point towards a low degree of awareness of the benefits of using CAT tools among non-adopters.

Also, the issue raised about the requirement of previous experience with CAT tools before adopting them, presented a significant difference between adopters and non-adopters of the tools. Only non-adopters perceived previous experience as a requirement for adopting CAT tools, which implies that inexperience with this type of technology and non-familiarity with these tools can represent a barrier towards adopting them.

Factor analysis was used to examine the differences in the perceptions of ICT in general and CAT tools in particular among the translators. A comparison of the factor analyses conducted for the translators' perceptions of ICT in general and towards CAT tools in particular revealed that in both cases, their perceptions could be grouped into three factors. While there was a common group of attitudes observed regarding "limitations" of ICT and CAT tools, there were two main differences in the structure of the perceptions of ICT and CAT tools. Firstly, the "benefits" and "problems" were perceived separately when looking at ICT in general, but they formed a single factor when looking at CAT tools in particular. Secondly, "experience" arose as a new factor for perceptions of CAT tools, treating it as a separate issue, rather than being considered as a problem, as was the case for the factor analysis for perceptions of ICT in general.

The comparison of the mean values of adopters and non-adopters of CAT tools and the assessment of their significance through ANOVA analysis revealed that the most significant difference was found in the perception on the "failure in using CAT tools" (c.f. 8.6.4). Results showed that CAT tool adopters did not consider their use of CAT tools a failure. Finally, the results of further comparisons within the different levels of experience with CAT tools also revealed a relationship between more successful use of CAT tools and more experience with them.

These findings seemed to indicate that there were important differences between the perceptions of the general use of ICT and the use of CAT tools. Although CAT tools are indeed one type of ICT that is used by translators, they were not seen as "just another software package" by the respondents. Benefits and problems derived from the use of ICT in general seem to be clearer to translators than the benefits and problems derived from CAT tools. Translators did not express clear benefits and problems of CAT tools, maybe because they are less familiar with the benefits and problems of these tools than of other ICT. In addition, as indicated by the issue separated by factor analysis from the rest, there seemed to be a major concern about having previous experience with CAT tools among freelancers in the sample that might affect the adoption of CAT tools.

8.6.1 Factor analysis on freelance translators' perceptions of ICT

In order to determine whether factor analysis was appropriate for the instrument measuring translators' perceptions of ICT, first, the adequacy of the correlations among the items of the instrument was examined. An initial inspection of the correlations revealed that 8 out of the 11 correlations were greater than 0.30 (i.e. they were statistically significant). This gave a first indication of the suitability of using factor analysis according to guidelines suggested for factor analysis adequacy (Hair et al., 1998, p. 99). Statistical tests such as KMO and Bartlett test of sphericity (very large value, 983.498, and very low significance, 0.000) were also used to confirm the overall factorability of the correlation matrix.[5] Another statistic which looks at the correlation of individual variables is the measure of sampling adequacy (MSA) in the anti-image correlation matrix. The measure of sampling adequacy was calculated for all of the variables and most of them were over 0.80, again falling in the "meritorious" range (Kaiser, 1970; Kaiser, 1974).

Factor analysis was then conducted using "principal components" analysis. This approach is concerned with determining the number of factors which account for the maximum amount of variance in the data (Hair et al., 1998). The latent root criterion suggested three factors since there were three eigenvalues greater than 1. Together these three factors represented 55.7% of the variability in the eleven variables. The scree test (Figure 8.2) also suggested the same number of factors should be retained for the subsequent analysis. The plot slopes steeply downwards from one factor to two factors, and more gently from two factors to three factors before becoming an approximately horizontal line. Although the first two factors were clearly differentiated from the rest, the third factor was also considered to be suitably differentiated from the rest. Also, using only the first two factors means that less than half of the variability (46.4%) is accounted for.

The unrotated component matrix with three factors was then orthogonally rotated with the varimax rotation method, the most widely used technique (Hair et al., 1998; Kellow, 2005). The varimax rotation criterion centres on simplifying the columns of the factor matrix and helps to make the pattern of the items associated with a given factor more distinct, thus increasing the interpretability of the factors (Kim, 1975).

The factor loadings and factor structure for all 11 items are presented in Table 8.7.

The results in Table 8.7 show that all the items measuring perceptions of ICT exhibited large factor loadings on at least one factor. As a rule of thumb, Hair et al. (1998) suggest that factor loadings greater than ±0.3 are considered to meet the minimal

---

[5] According to Stewart (1981), a value of 0.816 indicated a "meritorious" adequacy and hence indicates that the data are appropriate for factor analysis.

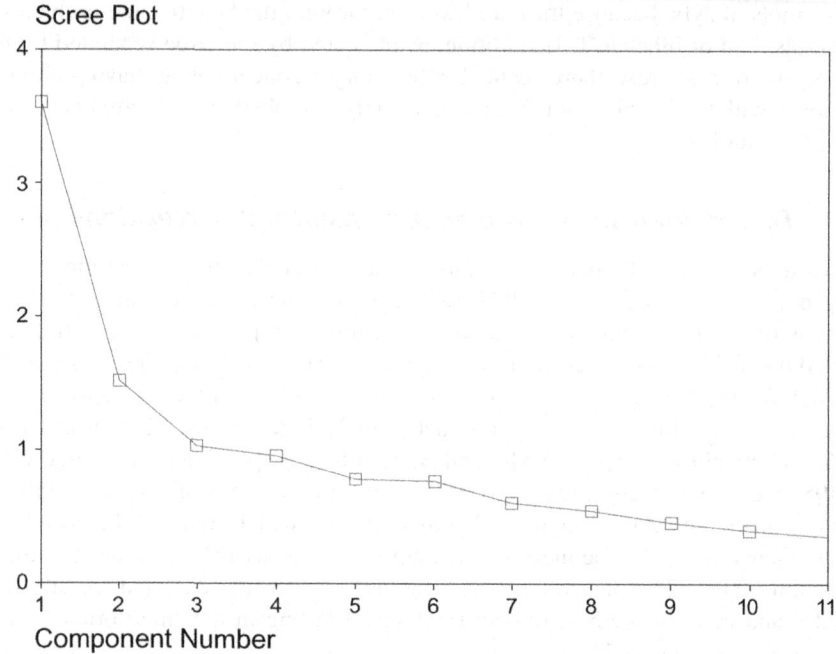

Figure 8.2 Scree test for factor analysis of perceptions of ICT

level, loadings of ±0.4 are considered more important, and if the loadings are ±0.5 or greater, they are considered very significant. In this case the results showed that all the items had a factor loading of more than 0.50 (the most demanding of the values suggested by Hair), implying that the items making up each of the factors were very significantly correlated to the factor itself. In fact, six of the variables had loadings greater than 0.7, which according to the guidelines given in Hair et al. (1998) would explain more than 50% of the variance in each of these variables.

Each of the items loaded significantly on exactly one of the three factors and an examination of the communalities (which represent the amount of variability accounted for by the factor solution) met the acceptable levels of explanation. Hence, the three-factor solution showed that there were three differentiated groups of items. These three groups could be summarised as: items showing benefits, items representing problems, and items showing limitations. The items linked to the first factor were considered to be items showing benefits. The five items allocated to this factor had loadings ranging from 0.623 to 0.777.

The items linked to the second factor were considered to be items representing problems. The four items allocated to this factor had loadings ranging from 0.558 to 0.707.

The items linked to the third factor were considered to be items showing limitations. The two remaining items allocated to this factor had loadings of 0.714 and 0.806. The presence of just two items loading on the last factor implies that a larger

Table 8.7 Factor analysis results for perceptions of ICT

Rotated Component Matrix[a]			
	Component		
	1	2	3
ICT: prev. exp. necessary	.453	.595	-1.675E-02
ICT: computerisation = higher quality services	.705	-.101	1.254E-02
ICT: computerisation = time saving benefits	.623	-.356	-7.512E-02
ICT: + benefits IF integrated apps	.152	-2.114E-02	.714
ICT: computerisation = + effectiveness as translator	.777	-.213	1.598E-02
ICT: computerisation = + revenue	.732	-.201	.102
ICT: computerisation = + comms with customers	.642	1.659E-02	-2.735E-02
ICT: apps failed to meet requirements	-.106	.146	.806
ICT: computerisation = many problems	-.196	.558	.340
ICT: use of apps = failure so far	-.334	.707	-3.743E-02
ICT: computerisation = - benefits than expected	-.471	.615	7.695E-02

Extraction Method: Principal Component Analysis.
Rotation Method: VarimaxwithKaiser Normalization.
[a] - Rotation conversed in 5 iterations.

number of questions on this area might have helped to define this factor more clearly. More questions on this area would probably have also helped to improve the measure of sampling adequacy in the anti-image test.

In summary, the results of factor analysis confirmed that there were three different types of perceptions of ICT, namely, benefits, problems and limitations.

A more detailed analysis of the "benefits factor" showed that the mean value for the items identified as benefits ranged from 4.06 to 4.33, which implied that most of the translators in the sample perceived all five items as important benefits of ICT. The highest mean value among the benefits was time saving, followed by improved communication with customers, and improved effectiveness as a translator. Providing higher quality services and increased revenue were perceived as slightly less important benefits, although they were still very significant. These findings suggest that translators in the sample considered their use of technologies as a way of improving their efficiency (time saving), their customer relations (communication with customers), and their productivity or quality levels (effectiveness as a translator). Nevertheless, the findings reveal that translators were also very concerned about "minor" benefits, such as meeting professional standards of quality through the use of ICT (higher quality of services) and obtaining better remuneration for their job (increase of revenue).

A detailed analysis of the "problems factor" showed that the mean values for the items identified as problems range from 4.00 to 1.91, which revealed a great deal of variation in the relative importance of the perceived problems relating to ICT. The need for previous experience with computers was seen as the most critical problem by most of the translators in the sample. The rest of the problems identified through factor analysis were perceived as rather less important, showing that overall their use of ICT had not been a failure, that the use of technologies had not brought fewer benefits than expected, and that computerisation had not created many problems. These findings suggest that, overall, translators in the sample were very concerned about having enough experience to cope with new ICT (previous experience with computers is necessary for adopting new applications), that they considered that they had succeeded in using technologies so far (use of applications has not been a failure so far), and that their use of computers and the technologies associated with them have been more positive than expected (gained fewer benefits than expected). The relative importance of the number of problems created by ICT (computerisation = many problems) was not very clear: although its mean value (2.84) revealed that ICT were not creating many problems, the standard deviation value (1.15) indicated a wide variation in the responses obtained. In other words, there could be some translators for whom ICT was creating few problems while there were other translators for whom it was creating many problems.

Finally, a closer look at the "limitations factor" showed that the mean value for the two items identified as limitations were 3.59 and 3.18. These mean values, along with standard deviation values close to 1 (0.94 and 1.06), implied that there was a wide diversity of positions with regard to the limitations of ICT. These findings suggested that there was not a majority of translators in the sample who considered that their use of technologies would bring them more or fewer benefits if their applications were more integrated (greater level of integration between applications). In a similar way, there was no majority of respondents who considered that the ICT they had used had matched or not their needs (applications failed to meet requirements). Since these two variables loaded onto the same factor, it also means that where translators see that their use of technologies would bring them more benefits if their applications were more integrated, they also find that applications are meeting their requirements (and vice versa). The scarcity of items that measured this factor was identified as a limitation of this research. More items need to be added to the instrument on perceptions of ICT, so that this aspect of translators' thinking could be more fully studied. Also, the fact that respondents gave a wide variety of responses means that there is scope for further research in this area. For example, why do some translators see that there would be benefits if their applications were more integrated, while other translators do not see this?

8.6.2 Factor analysis on the perceptions of CAT tools

For comparison purposes, the 11-item instrument used to measure perceptions of ICT was adapted to measure translators' perceptions of CAT tools. The statements in the original instrument were worded slightly differently to capture translators' perceptions of CAT tools. One of the items ("Computerisation significantly improves my communication with customers") was not applicable to the use of CAT tools, and was substituted by an item asking about the respondents' opinion on the cost of CAT tools (i.e. "CAT tools are well worth their cost"). This issue of costs was one that arose from the literature examined on translators' opinions about CAT tools. Factor analysis was again performed in order to achieve a better understanding of the structure of the data and to identify underlying dimensions.

Again, adequacy of the correlations among the items of the instrument was examined against the suitability of using factor analysis. An initial inspection of the correlations revealed that all the correlations but two (9 out of the 11) were greater than 0.30 (i.e. they were statistically significant). The statistical tests KMO (value of 0.871 indicated a "meritorious" adequacy) and Bartlett test of sphericity (very large value, 1771.253, and very low significance, 0.000) were also used to confirm the overall factorability of the correlation matrix.

Factor analysis was again conducted using principal components analysis and again three eigenvalues were more than 1. However, in this case, the three factors with eigenvalues over 1 explained a larger proportion of the variability of the eleven variables: 68.7%. The scree test also suggested the same number of factors should be retained for the subsequent analysis (Figure 8.3). Again, the plot shows a very clear first factor and then slopes gently from two factors to three factors before becoming an approximately horizontal line. Although the first and second factors were more clearly differentiated from the rest, the third factor was also considered to be suitably differentiated from the rest. Also, using the same number of factors that were used in examining the perceptions of ICT in general would aid comparability when considering the perceptions of this particular type of ICT (CAT tools).

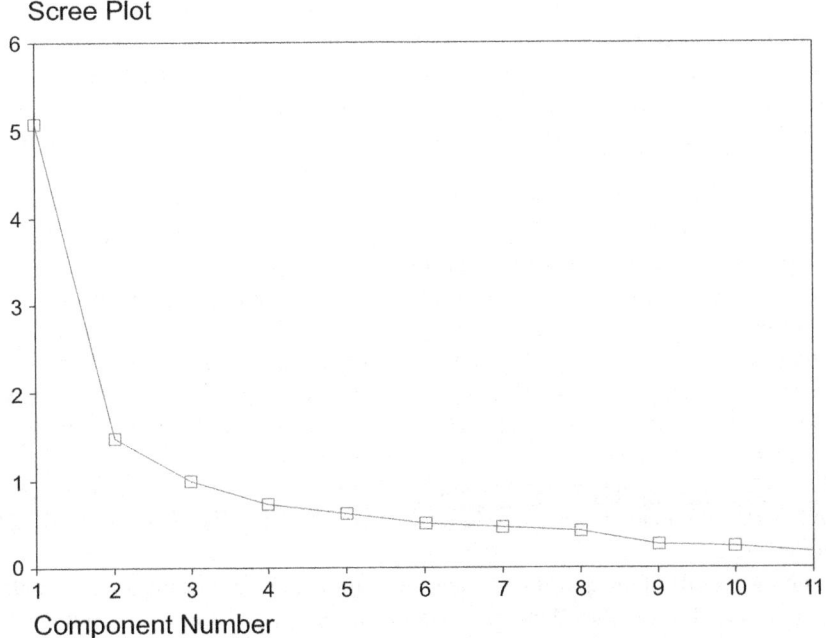

Figure 8.3 Scree test

Table 8.8 Factor analysis results for perceptions of CAT tools

Rotated Component Matrix[a]			
	Component		
	1	2	3
CAT: prev. exp. necessary	-1.726E-02	.908	-6.343E-02
CAT = higher quality services	.808	-5.546E-02	2.670E-02
CAT = time saving benefits	.810	-.173	9.081E-02
CAT are worth their cost	.767	-.131	-.171
CAT = + revenue	.877	-7.996E-02	-1.330E-02
CAT = + effectiveness as translator	.860	-.165	4.282E-03
CAT = + effectiveness as translator IF integrated	.412	-.138	.590
CAT failed to meet requirements	-.120	4.979E-02	.845
CAT = many problems	-.463	.412	.425
CAT = failure so far	-.575	.556	.124
CAT = - benefits lhan expected	-.594	.403	.455

Exiraction Method: Principal Component Analysis.
Rotation Melhod: Varimax with Kaiser Normalization.
[a] - Rotation converged in 5 iterations.

The unrotated component matrix with three factors was then orthogonally rotated with the varimax rotation method. All 11 items measuring perceptions of CAT tools exhibited large factor loadings on at least one factor (Table 8.8).

Each of the items loaded significantly on exactly one of the three factors, except for the item about the problems created by CAT tools, which showed similar correlations with all three factors (ranging from 0.412 to 0.463). However, the examination of the communalities of the factor solution met the acceptable levels of explanation. The smallest communalities were slightly above 0.5 and were for the item about the problems created by CAT tools and the item about the integration of CAT tools.

One possible way to improve the clarity of the factor solution is to try a different number of factors, which, in this case, did not prove to help much. Another possible way to improve the clarity of the factor solution is to try a different rotation method. In particular, it was thought that a different orthogonal rotation method might be useful. In order to try to bring clarity to the allocation of the item on problems created by CAT tools, the factor analysis was re-run using quartimax rotation. Comparing the results obtained using quartimax rotation with the results obtained using varimax rotation, it was found that overall they were very similar. However, the item on problems created by CAT tools now loaded significantly (0.577) only on the first factor, i.e. "Benefits/ Problems."

Using the results of the quartimax rotation to allocate variables to factors, the three-factor solution showed that there were three differentiated groups of items: items showing benefits and problems, one item showing experience and items showing limitations. With regard to the "benefits/problems" factor, eight items were allocated to this factor with loadings ranging from 0.575 to 0.877. Only one item loaded on the factor representing "experience" with a factor loading of 0.908. Finally the two remaining items loaded on the "limitations" factor with loadings of 0.590 and 0.845.

Overall these results were disappointing; in factor analysis it is expected that there will be several questions loading on each factor. Clearly, the factor structure apparent with general ICT did not simply carry over to the specific application (CAT tools). Further research was needed to identify underlying dimensions (types of perceptions) in translators thinking about CAT tools.

A more detailed analysis of the "benefits/problems" factor showed that the mean values for the items identified as benefits/problems ranged from 2.64 to 3.44, which is lower than the range of mean values for the ICT benefits (4.06 to 4.33). This suggested that most of the translators in the sample did not have a clear perception of the benefits and problems related to CAT tools. This variation in responses might be due to differences between the perceptions of translators who had adopted CAT tools and those who were not using them. The highest mean value among the benefits/problems was time saving, followed in descending order by providing higher quality services, increased revenue, improved effectiveness as a translator, the origin of CAT tool problems, perceptions of the value of CAT tools, failure in using CAT tools and obtaining fewer benefits than expected. These findings suggest that translators in the sample considered CAT tools to be a way of improving their efficiency (time saving), meeting professional quality standards (higher quality of services), obtaining a better remuneration for their job (increase of revenue), and increasing their production and quality levels (effectiveness as a translator). Since these eight variables loaded onto the same factor, it also meant that (for example) when translators see CAT tools as saving time they also see CAT tools as providing higher quality services and improving their effectiveness as a translator (and vice versa). There was no evidence here that CAT tools require a trade-off between cost (as measured by saving time) and quality.

The only item loading significantly on the "experience factor" arose as a separate factor with a very significant loading (0.908). Despite this being the only item loading very highly on this factor, one of the items loading on the benefits/problems factor (the use of CAT had been a failure so far) also loaded significantly on the experience factor (0.556). This suggested that the experience needed to use CAT tools was a clear concern among the translators in the sample. While it was disappointing to have a factor with only one item loading on it, this was an important factor. However, the scarcity of items that measured this factor was identified as a limitation of the study.

A closer look at the "limitations factor" showed that the mean values for the two items identified as limitations were 3.30 and 3.20. These findings suggested that there was some variation in responses when translators evaluated the limitations of CAT tools. Also, this implied that there was not a majority of translators in the sample who considered that their use of technologies would bring them more or fewer benefits if their applications were more integrated (greater level of integration between applications). The lack of clarity obtained from this factor, the limitations of CAT tools, along with the scarcity of items that measured this factor were identified as a limitation of this research that was addressed in the second phase of the study (c.f. section 8.5.6).

Table 8.9 **A comparison of ICT and CAT tool factors**

Factors examining perceptions	Towards use of ICT in general	Towards use of CAT tools
Factor 1	Benefits	Benefits / Problems
Factor 2	Problems	Experience
Factor 3	Limitations	Limitations

8.6.3 A comparison of factor analysis results for perceptions of CAT tools and ICT in general

The results of factor analysis conducted to examine the attitudes that translators in the sample had towards ICT in general and towards CAT tools in particular revealed that there were important differences between the perceptions of the general use of ICT and the use of a specific type of ICT (CAT tools). Although CAT tools are indeed one type of ICT that is used by translators, it was not seen as "just another software package" by the respondents. Comparing the results of factor analysis for ICT and for CAT tools, there were a number of similarities and differences between the items grouped for each factor (Table 8.9).

There were two main differences observed in the grouping of the items, namely, the allocation of items relating to the "problems" factor in both analyses, and the rise of a separate factor regarding the experience needed to use CAT tools.

When talking about ICT in general, benefits and problems were perceived as two different factors; however, when translators gave their opinions on CAT tools, items looking at benefits (or positive effects) and problems (or negative effects) were combined into the same factor. This suggested that the benefits perceived to be gained from ICT in general are clearly differentiated in the minds of translators from the problems they cause. All the benefits were loaded together as a first factor since they are seen to be strongly linked to each other, while all the perceived problems were loaded together as a second factor. On the other hand, the benefits from CAT tools were not differentiated in the minds of translators from the problems they cause. This might be because translators were less familiar with the benefits and problems of CAT tools. Alternatively, the difference might be inherent in the difference between "general" and "specific."

At a detailed level, all of the items in the factor "benefits of using ICT" also appeared in the factor "benefits/problems of using CAT" (apart from the question about communications which was not one of the items for the CAT tools). Similarly, all of the items in the factor "problems of using ICT" also appeared in the factor "benefit/problems of using CAT" apart from the question about experience. In other words, the issues which were clearly separated in the minds of translators when asking about ICT in general, were joined together when asking about CAT in particular.

The other big difference concerns the item about previous experience (part of the "Problems" factor with a loading of 0.595), which also had a rather significant loading on the "Benefits" factor (0.453). This showed that previous experience was mostly linked to other problems in the minds of translators but also possibly had some links with benefits in the minds of the translators. This possible ambiguity in the perception of the need for previous experience disappeared when the analysis turned towards the perception of CAT tools, where previous experience appeared as a separate factor.

Again at a detailed level, the factors identified as "limitations in using ICT" and as "limitations in using CAT tools" included the same items in both cases, i.e. that the translator's effectiveness would be increased if there was a higher degree of integration within the applications they use, and that ICT and CAT tools had failed to meet their requirements. It seemed, therefore, that the links between questions about limitations perceived for ICT in general also applied to the case of CAT tools.

8.6.4 A comparison of CAT tool perceptions of adopters and non-adopters

In order to examine how perceptions of CAT tools differed between adopters and non-adopters, their mean values were compared and their statistical significance assessed using ANOVA at the 0.05 significance level (Table 8.10).

Table 8.10 CAT perceptions of adopters and non-adopters

	Mean difference between groups of respondents based on CAT use			
	Mean Scores		F-Test (ANOVA)	
Perceptions on CAT tools	*Group 1: CAT adopters*	*Group 2: CAT non-adopters*	*F Ratio*	*Significance*
Previous experience with CAT is necessary	2.68	3.15	18.379	0.000
CAT = higher quality services	3.84	2.97	97.233	0.000
CAT = time saving benefits	4.03	3.21	87.172	0.000
CAT are worth their cost	3.47	2.87	44.380	0.000
CAT = + revenue	3.74	3.00	76.953	0.000
CAT = + effectiveness as translator	3.82	2.83	114.372	0.000
CAT = + effectiveness as translator IF integrated	3.61	3.22	17.301	0.000
CAT failed to meet requirements	3.38	3.16	5.554	0.019
CAT = many problems	2.73	3.17	22.637	0.000
CAT = failure so far	1.71	3.03	195.909	0.000
CAT = - benefits than expected	2.46	3.11	35.041	0.000

All the perceptions were found to be statistically very significant at the 0.05 level (all the variables had a 0.000 significance, except for "CAT failed to meet requirements" which had a p-value of 0.019). The variable which showed the largest difference was "CAT = failure so far," with an F value of 195.909. Here the mean value for adopters (1.71) was much lower than the mean value for non-adopters (3.03), and this indicated a clear difference between the perceptions in each group. In particular, the adopters largely disagreed with the statement formulated for this variable, showing that their use of CAT tools had not been a failure.

Other variables that were found to be very significant were "CAT = higher quality services" ($F = 97.233$), "CAT = time saving benefits" ($F = 87.172$), and "CAT = + revenue" ($F = 76.953$) with mean values from the adopters group close to "4 = Agree" (3.84, 4.03 and 3.74, respectively). This implied that adopters claimed that CAT tools helped them offer higher quality services, that they brought time savings, and that the use of CAT tools increased their revenue. Again it is important to note that there was no evidence that CAT tools require a trade-off between cost (as measured by saving time) and quality.

As would be expected from those respondents who had not adopted CAT tools so far, all the perceptions of the members of this group (Group 2 in the table) were close to the middle value of the 5-point Likert scale ranging from "1 = Strongly Disagree" to "5 = Strongly Agree." The middle value was labelled as "Don't Know."

8.6.5 Level of CAT experience and CAT perceptions

ANOVA at the 0.05 significance level was again used to examine how perceptions of CAT tools differed within different levels of experience with CAT tools among the respondents. Their mean values were compared and their statistical significance assessed using the mean values of the responses to the perceptions of CAT tools across respondents with different levels of experience (Table 8.11).

As would be expected from those respondents who were not familiar with CAT tools, all the perceptions of the members of this group (Group 1 in the table) were close to the middle value (which was labelled as "Don't Know") of the 5-point Likert scale ranging from "1 = Strongly Disagree" to "5 = Strongly Agree."

All the perceptions (for all groups) were found to be statistically very significant at the 0.05 level (all the variables had a 0.000 significance, except for "Previous experience with CAT is necessary," which had a p-value of 0.001, and "CAT = many problems," which had a p-value of 0.002). The variable which showed the largest difference was "CAT = failure so far," with an F value of 61.533. Here the mean value for the group with "extensive experience with CAT tools" had a mean of 1.47, which was much lower than the means for the other groups. The progression in the mean values showed that the more experienced respondents were with CAT tools, the more success they had with this technology. However, it is recognised that this result might be "self-fulfilling" in the sense that if CAT was a failure for some translators then these translators would not persevere to achieve "extensive experience." On the other hand, none of the groups showed signs of failure in using CAT tools (i.e. the mean values for respondents in all experience groups disagreed with the statement "My use of CAT tools has been a failure").

Table 8.11 Perceptions of respondents with different levels of experience with CAT tools

	Mean difference between groups of respondents based on CAT experience					
	Mean Scores				F-Test (ANOVA)	
Perceptions on CAT tools	Group 1: Not familiar	Group 2: Familiar with no experience	Group 3: Familiar with some experience	Group 4: Familiar with extensive experience	F Ratio	Sig.
Previous experience with CAT is necessary	3.14	3.20	2.84	2.67	5.743	0.001
CAT = higher quality services	2.95	3.01	3.30	3.95	27.509	0.000
CAT = time saving benefits	3.08	3.33	3.59	4.18	34.368	0.000
CAT are worth their cost	2.93	2.88	2.75	3.73	25.185	0.000
CAT = + revenue	3.00	3.05	3.09	3.97	32.054	0.000
CAT = + effectiveness as translator	2.83	2.81	3.34	3.18	47.634	0.000
CAT = + effectiveness as translator IF integrated	3.09	3.34	3.56	3.60	8.201	0.000
CAT failed to meet requirements	3.04	3.18	3.57	3.33	6.698	0.000
CAT = many problems	3.09	3.18	3.13	2.71	5.059	0.002
CAT = failure so far	3.03	2.99	2.58	1.47	61.533	0.000
CAT = - benefits than expected	3.04	3.08	3.32	2.08	26.233	0.000

Other variables that were found to be very significant again included "CAT = higher quality services" ($F = 27.509$), "CAT = time saving benefits" ($F = 34.368$) and "CAT = + effectiveness as translator" ($F = 47.634$) with mean values from the most experienced group near or above "4 = Agree" (3.95, 4.18 and 4.10 respectively). This implied that those respondents with more experience with CAT tools were the ones who saved more time and who had seen their effectiveness increase. Also, this again implied that there is no evidence that CAT tools require a trade-off between cost (as measured by saving time) and quality (including effectiveness as a translator).

8.6.6 Technology attributes affecting CAT tool adoption

In addition to determining the organisational characteristics that affect CAT tool adoption (i.e. the characteristics of the freelance translators, the translation business and translators' perceptions of ICT), the identification of the factors that motivate or inhibit the adoption has previously helped to achieve a better understanding of how ICT adoption is affected in the IS domain. The qualitative analysis of the data collected during the second phase of the study provided important insights about the factors that had a positive effect on the decision to adopt CAT tools (motivators), and what factors had a negative effect on this decision (inhibitors).

The instruments used to measure the perceptions are described in Chapter 7 and were based on Moore and Benbasat's constructs to measure the perceptions that an individual may have of adopting an IT innovation (Moore and Benbasat, 1991, p. 192); namely, relative advantage, compatibility, voluntariness, image, ease of use, result demonstrability, visibility and trialability (c.f. section 7.4.2.1.). Figures 8.4 and 8.5 present the summarised data of the two descriptive meta-matrix displays created by the data reduction process for each translator according to the constructs being studied and stacked according to their different contexts (CAT adoption or CAT non-adoption).

ID	CAT TOOL USED	READ	COMP	EASU	VOLU	IMAG	VISI	TRIA	REDE	Main subject areas*	Main type of client
ad01	Trados SDLX Transit	5.0	5.0	4.3	4.5	3.0	3.0	2.0	4.3	1, 2, 4	Translation agencies
ad11	Trados	5.0	5.0	3.8	1.5	2.8	4.0	4.0	4.0	1, 3	Translation agencies
ad15	Wordfast	5.0	5.0	5.0	4.0	2.5	4.0	5.0	5.0	1, 2, 4	Translation agencies
ad17	Trados	5.0	4.0	3.8	1.5	3.0	3.0	1.0	3.7	1, 2	Translation agencies
ad18	Déjà Vu	5.0	4.5	3.5	4.5	2.0	4.5	4.3	5.0	1, 5	Direct clients
ad19	Trados	5.0	4.5	3.0	3.0	3.8	2.5	1.0	4.0	1, 6	Translation agencies
ad07	Déjà Vu	4.5	3.0	3.5	5.0	2.8	3.5	5.0	4.0	1, 3	Translation agencies
ad03	LionLinguist Trados WordFast	4.3	3.0	2.5	2.0	4.0	3.5	1.3	4.0	1, 6	Direct clients
ad14	Star Transit	4.3	4.0	4.0	3.5	3.3	4.0	4.0	4.3	1, 3	Half agencies, half direct
ad02	Déjà Vu & SDLX before	4.0	4.0	3.5	4.0	3.0	2.5	1.7	4.0	1	Translation agencies
ad10	Wordfast	4.0	4.0	3.2	3.0	4.0	4.0	3.3	4.0	1, 2, 6	Translation agencies
ad16	Trados & Star Transit	4.0	4.0	4.0	2.5	3.8	3.5	2.0	4.0	1, 2	Translation agencies
ad04	Trados	3.9	4.5	4.2	2.5	3.3	3.0	1.3	2.7	1, 2	Translation agencies
ad05	Trados	3.9	4.0	3.7	3.0	4.0	1.5	1.0	4.0	2	Translation agencies
ad06	Trados	3.9	4.0	2.5	3.5	3.8	2.5	1.0	3.7	2, 4, 5	Translation agencies
ad12	Trados	3.8	4.0	3.0	2.5	3.0	3.0	2.0	4.0	2	Translation agencies
ad13	Wordfast	3.8	4.0	3.8	4.0	2.3	3.0	4.0	4.0	1, 2	Translation agencies
ad09	Trados	3.4	3.5	3.8	3.0	4.8	3.0	2.7	4.0	1, 6	Translation agencies
ad08	Déjà Vu	3.3	3.5	3.8	4.5	3.8	3.0	4.0	4.0	1, 2	Translation agencies

*Subject areas: 1 = Business/Commerce; 2 = Technical; 3 = Legal; 4 = Medical; 5 = Scientific; 6 = Arts/Tourism
Key for constructs:READ = relative advantage; COMP = compatibility; EASU = ease of use; VOLU = voluntariness; IMAG = image; VISI = visibility; TRIA = trialability; and REDE = result demonstrability.

Figure 8.4 Descriptive matrix of the factors affecting CAT tool adoption among adopters

ID	READ	COMP	EASU	VOLU	IMAG	VISI	TRIA	REDE	Main subject areas*	Main type of client
na01	3.1	2.0	3.0	5.0	3.0	2.0	3.0	3.0	1, 3	Direct clients
na02	4.3	3.5	4.3	5.0	3.5	3.5	3.3	4.3	6	Translation agencies
na03	3.4	2.0	2.2	2.0	3.0	3.0	3.5	2.3	1, 3, 4	Translation agencies
na04	3.0	1.0	1.8	4.0	3.3	3.5	3.5	3.7	1, 5	Direct clients
na05	4.0	4.0	4.5	5.0	1.0	2.0	3.0	4.0	1, 3	Translation agencies
na06	2.5	2.0	2.7	5.0	2.5	3.5	4.0	2.7	1, 3	Translation agencies
na07	2.6	2.5	2.3	3.0	3.0	2.0	2.5	1.7	1, 3, 6	Direct clients
na08	1.1	2.5	2.8	1.0	5.0	5.0	3.3	2.3	6	Translation agencies
na09	1.0	1.0	2.3	1.0	4.0	4.0	3.0	3.7	4	Translation agencies
na10	3.3	3.0	4.0	3.0	2.3	4.0	3.0	4.0	6	Half agencies, half direct
...	:	:	:	:	:	:	:	:	:	:
na30	4.0	4.0	3.7	2.0	4.5	4.0	2.0	3.3	3, 6	Direct clients
na31	3.0	3.5	3.5	4.0	3.5	2.0	2.8	3.3	3, 6	Direct clients
na32	3.4	3.0	3.3	5.0	3.3	3.5	3.0	3.3	3, 6	Direct clients
na33	4.0	4.0	2.3	4.0	1.5	2.5	1.0	3.7	1	Translation agencies
na34	2.3	2.0	2.3	4.0	2.8	5.0	4.5	4.5	1, 6	Direct clients

*Subject areas: 1 = Business/Commerce; 2 = Technical; 3 = Legal; 4 = Medical; 5 = Scientific; 6 = Arts/Tourism

Key for constructs: READ = relative advantage; COMP = compatibility; EASU = ease of use; VOLU = voluntariness; IMAG = image; VISI = visibility; TRIA = trialability; and REDE = result demonstrability.

Figure 8.5 Descriptive matrix of the factors affecting CAT tool adoption among non-adopters

After this first exploratory step of the cross-case analysis to see what the cases look like and how they fit within each defined context of adoption, Miles and Huberman's framework suggests going beyond the description of multiple cases to generate explanations and to test them systematically, which is considered by the authors as "our best resource for advancing our theories about the way the world works" (Miles and Huberman, 1994, p. 207). Thus, the analysis moved from the initial matrices used for reducing, organising and describing the data, to a more detailed level of analysis which allowed us to start inferring conclusions from the matrices. Again following Miles and Huberman's recommendations, *case-ordered predictor-outcome matrices* were used. Using this method, the adoption of CAT tools (i.e. whether the translator was an adopter or not) was used as the main criterion/outcome variable. Then, the adopter and non-adopter cases of the descriptive matrices were arrayed in new matrices, one for each of the constructs that were considered predictors of the adoption of CAT tools (i.e. the main antecedent variables thought to be most important contributors to the outcome). Due to a number of reasons, such as the flexibility for working with text and figures, and the ease of use for entering, processing, displaying and arranging data, spreadsheet software (Microsoft Excel) was used as the canvas and tool to build tables and matrices. More specific questions were formulated to account for the factors that were affecting the adoption of CAT tools (i.e. what Miles and Huberman called "asking prediction questions"). A number of predictors were then used to observe what was going on in all the cases of each setting at the same time, using the summarised construct variables as predictors, and some of them (READ, COMP, EASU, TRIA and REDE) as criteria to display the ordered cases within each setting. The conceptual and empirical considerations for using these predictors as criteria to conduct the cross-case analysis stemmed from the issues arising from the existing literature, and from the findings of the first phase of the present research:

- The ambiguity in the benefits and problems derived from the use of CAT tools observed in the translators during the first phase of this research, along with the existing scepticism on the advantages provided by the use of CAT tools, made the READ (Relative Advantage) construct an important criterion predictor to analyse the multiple cases of this phase of the research.
- The existing concern about the compatibility of CAT tools with translators' way of working, or about the suitability of these tools for certain types of translation jobs, made the COMP (Compatibility) construct an important criterion predictor to analyse the multiple cases of this phase of the research.
- The evidence of a concern about the need of previous experience with CAT tools before adopting them was related with the issues examined through the EASU (Ease of Use), TRIA (Trialability) and REDE (Result Demonstrability) constructs, which made them important criteria predictors to analyse the multiple cases of this phase of the research.

As a result of ordering the cases according to their degree of importance of the criteria predictors, the values of the predictors represented three degrees of importance: "low" (values between 1 and 2.3 on the scale), "moderate" (values between 2.3 and 3.6), and "high" (values over 3.6). In addition, the relative importance of the items under the predictor variables could be calculated by finding out the percentage of respondents who agreed with each item (i.e. values 4 or 5 on the scales) under each variable. With the scaled predictor variables and the cases from the previous two descriptive matrices (for adopters and non-adopters), the next step involved the construction of the predictor-outcome matrices. Five matrices were created, including both CAT tool adopter and non-adopter cases and ordering them according to each of the predictor variables selected as criteria for examining the cases. The data from the scales were represented in their numeric value, showing the summarised value of each construct, placing the criteria predictor in first place, and identifying the three degrees of importance by different degrees of shading (dark grey for "high importance," light grey for "moderate importance," and white for "low importance"). In addition, values that were considered highly important (i.e. higher than 3.6) were shown in bold font face. Shading was also used in the "id labels" to mark adopter cases (light shading) and non-adopter cases (dark shading). This shading format of the matrices was helpful in spotting the different degrees of importance of the predictor variables across the cases, and in the process of drawing patterns, which helped to draw conclusions. A sample of part of one of the matrices created is presented in Figure 8.6.

Conclusions were drawn from the predictions by using the Miles and Huberman method detailed in Chapter 7 (c.f. 7.5.2.). The specific tactic used is indicated between squared brackets at the end of each conclusion.

- *Relative Advantage* (READ) was perceived as the most important characteristic of adopting CAT tools. In particular, enhancing the translators' effectiveness, making their job easier, and increasing their job performance and productivity were the most important advantages gained by the adopters of CAT tools [*noting patterns, counting, making contrasts/comparisons, noting relations between variables*].
- *Compatibility* (COMP) also proved to be an important characteristic of adopting CAT tools. The relative importance of the compatibility of the translators' way of working and the type of assignments they undertook was very high among those translators who had adopted CAT tools, with some prevalence of the compatibility with their working style over the type of

COMP	CASE ID	READ	EASU	VOLU	IMAG	VISI	TRIA	REDE
5.0	ad01	5.0	4.3	4.5	3.0	3.0	2.0	4.3
5.0	ad11	5.0	3.8	1.5	2.8	4.0	4.0	4.0
5.0	ad15	5.0	5.0	4.0	2.5	4.0	5.0	5.0
4.5	ad18	5.0	3.5	4.5	2.0	4.5	4.3	5.0
4.5	ad19	5.0	3.0	3.0	3.8	2.5	1.0	4.0
4.5	na21	4.0	2.7	4.0	2.0	1.5	3.8	3.0
4.5	ad04	3.9	4.2	2.5	3.3	3.0	1.3	2.7
4.0	ad17	5.0	3.8	1.5	3.0	3.0	1.0	3.7
4.0	ad14	4.3	4.0	3.5	3.3	4.0	4.0	4.3
4.0	ad02	4.0	3.5	4.0	3.0	2.5	1.7	4.0
4.0	ad10	4.0	3.2	3.0	4.0	4.0	3.3	4.0
4.0	ad16	4.0	4.0	2.5	3.8	3.5	2.0	4.0
4.0	na05	4.0	4.5	5.0	1.0	2.0	3.0	4.0
4.0	na30	4.0	3.7	2.0	4.5	4.0	2.0	3.3
4.0	na33	4.0	2.3	4.0	1.5	2.5	1.0	3.7
4.0	ad05	3.9	3.7	3.0	4.0	1.5	1.0	4.0
4.0	ad06	3.9	2.5	3.5	3.8	2.5	1.0	3.7
4.0	ad12	3.8	3.0	2.5	3.0	3.0	2.0	4.0
4.0	ad13	3.8	3.8	4.0	2.3	3.0	4.0	4.0
4.0	na25	3.0	2.2	4.0	3.8	2.5	3.0	3.3
3.5	na02	4.3	4.3	5.0	3.5	3.5	3.3	4.3
3.5	na20	3.6	2.8	4.0	3.0	1.5	2.3	3.7
3.5	na14	3.5	3.0	4.0	3.8	2.0	3.3	3.0
3.5	ad09	3.4	3.8	3.0	4.8	3.0	2.7	4.0
3.5	ad08	3.3	3.8	4.5	3.8	3.0	4.0	4.0
3.5	na31	3.0	3.5	4.0	3.5	2.0	2.8	3.3
3.0	ad07	4.5	3.5	5.0	2.8	3.5	5.0	4.0

Figure 8.6 Predictor-outcome matrix of relevance of the construct "Compatibility" (COMP) for CAT tool adoption

assignments they dealt with. The relative importance of COMP in the non-adoption setting was, on the other hand, quite negative, indicating that translators who had not adopted CAT tools did not perceive this software as compatible with their translation job [*noting patterns, counting, making contrasts/comparisons*].

- More predicting variables related to the translators' way of working and their type of translation work were examined to further explain compatibility of CAT tools, such as the subject areas in which they were working or the main type of client they were dealing with. Analysis showed that among the adopters, the use of CAT tools was especially compatible with those translators working in the technical, scientific and business/commerce subject areas, while CAT tools were not particularly compatible with those working in legal translation. Among the non-adopters of CAT tools, the most usual subject areas of work were legal, business/commercial and arts/tourist/literary translation. With regard to the types of client, the adopters of CAT tools were mostly working for translation agencies, while the non-adopters worked in a similar proportion for agencies and for direct clients. For this reason, the compatibility predictor showed a relationship with translators whose type of client was mainly agencies [*counting, making contrasts/comparisons*].
- *Ease of use* (EASU) was presented as a positive perception of adopting CAT tools by adopters. However, there was a tendency towards a negative perception of adopting CAT tools due to the low EASU observed among the non-adopters. This meant that not finding it easy to

learn and use CAT tools could affect the adoption of CAT tools negatively [*noting patterns, counting, making contrasts/comparisons*].
- Having more or fewer opportunities to try CAT tools out before adopting them was not found to be a determinant characteristic of the tools for the adoption process [*noting patterns, counting, making contrasts/comparisons*].

These conclusions were further tested and predictions were strengthened following Miles and Huberman's advice (1994, p. 310). Relationships between the predictors were tested by comparing the progression of their numeric scales, thus helping to strengthen the prediction that these constructs were the ones affecting the adoption of CAT tools more positively. In order to understand how the specific perceptions of adopting CAT tools affected their adoption by translators, the relative importance of the particular issues investigated under the main key constructs was analysed.

A linear progression of the cases ordered by each relevant construct was used to test if each item/factor within the constructs was pervasive and significant, and therefore a frequency measure was used, instead of conventional averaging. In addition, this level of detail helped the analysis to confirm the conclusions drawn from the whole picture provided by the matrices at a micro level where nuances were defined by the cross-tabulation of constructs and the number of cases supporting each value. As a result, for example, the perceived advantages gained from CAT tool adoption were ranked according to their relative importance (Table 8.12).

In summary, the findings of this part of the study revealed a number of factors that were likely to motivate CAT tool adoption, such as gaining a relative advantage from the use of CAT tools, the compatibility of CAT tools with the type of work undertaken by the adopters, and the communication of the advantages and disadvantages of using CAT tools by the adopters of these tools to others. On the other hand, the main factor found to be inhibiting CAT tool adoption was the fear of learning and using CAT tools among non-adopters. Also, some factors which were not found to be significant for the adoption of CAT tools were the voluntariness for using CAT tools, the opportunity

Table 8.12 **Perceived relative advantages associated with CAT tool adoption**

Rank	Score*	Relative advantage
1	94.7**	CAT tools are overall advantageous
2	94.7**	CAT tools enhance effectiveness
3	94.7**	CAT tools make job easier
4	94.7**	CAT tools improve job performance
5	89.5	CAT tools increase productivity
6	84.2	CAT tools improve the quality of the work
7	78.9	CAT tools enable to accomplish tasks more quickly
8	63.2	CAT tools give more control over work

* Score for a relative advantage represents the percentage of respondents who rated the advantage as either ageeing or strongly agreeing with the statement rating it.
** Four of the scores were equal to 94.7, so those with a higher percentage of respondents strongly agreeing (5) on the advantage were ranked first.

to try the tools before deciding to adopt them, the consequences of CAT tool usage on the translator's image and observing other translators using CAT tools.

8.7 Impacts of CAT tool adoption

One of the concerns behind the scepticism among translators about CAT tools was whether the tools can really deliver what they promise, essentially higher productivity and improved quality of their translations (see, for example, Heyn, 1998; Somers, 2003b). In addition, the lack of research about CAT tool adoption by freelance translators also meant that there was little or no evidence of the impacts that CAT tools have on freelance translators that could support the claims made about their benefits.

The responses obtained from the instrument measuring impacts of CAT tool adoption – for adopters – and perceived impacts – for non-adopters – (c.f. 7.4.2.1) were classified as *positive* effects of CAT tool adoption (e.g. a small/large increase in the translator's turnover due to the use of CAT tools), as *neutral* (e.g. the translator's productivity remained unchanged while using CAT tools), or as *negative* effects (e.g. the prices charged by the translator decreased due to the use of CAT tools).

In order to determine the relative importance of the impacts of CAT tools among adopters, the data from the cases in the matrix indicating positive and negative impacts of CAT tools on translators' work were ranked according to the percentage of translators reporting positive impacts of CAT tools. Therefore, the percentage of adopters who stated they experienced a small or large increase in each of the impacts was calculated (Table 8.13).

As reported by the summarised findings above, the study provided empirical evidence of the impacts of CAT tool adoption. They were largely positive, the most important impacts being an increase in the quality of their translations undertaken and an increase in their productivity. The only negative impact detected, in only a few cases, was a slight decrease in the prices charged because clients might want to pay less for

Table 8.13 **Positive impacts of adopting CAT tools (adopters)**

Rank	Score*	Impact on...
1	89.5%	Quality of translations
2	88.9%	Translator's productivity
3	72.2%	Volume of work undertaken
4	66.7%	Volume of work offered to translators by clients
5	61.1%	Translator's turnover
6	42.1%	Size of translator's customer base
7	42.1%	Number of clients
8	5.6%	Prices translators charge for their work

* Score for an impact having a positive effect represents the percentage of respondents who rated CAT tools as causing some or a large increase in the impact in the statement.

reutilising previous translations. The two main impacts found in this study, therefore, confirmed the two main benefits attributed to the use of CAT tools: increased productivity and increased quality of work.

Non-adopters of CAT tools were also asked about the specific impacts that they believed this specialised ICT would have on their work for contrast and comparison purposes with adopters' findings. The importance of the beliefs that the non-adopters of CAT tools had with regard to the impacts of using them was examined in a similar way to how it was done with adopters. The data from the cases in the non-adopter matrix indicating the impacts that translators thought CAT tools would have on their work were ranked according to the percentage of translators reporting positive impacts of CAT tools. Therefore, the percentage of adopters who thought they would experience a small or large increase in each of the impacts was calculated, originating both positive and negative perceived impacts.

Overall, non-adopters perceived that impacts were not as positive as the impacts claimed by the current adopters of the specialised ICT (Table 8.14). The major positive impacts highlighted by adopters of CAT tools (increased quality of translations and increased productivity) only ranked 7th and 5th respectively from non-adopters' views, and in both cases, with less than half of the respondents thinking that CAT tools would have a positive impact on these issues. In addition, the main positive impacts perceived by non-adopters were increasing the volume of work offered to translators by clients, increasing translators' turnover, and increasing the size of translators' customer base. All three cases were among the less important impacts detected by current adopters of CAT tools. The only common negative impact between adopters and non-adopters of CAT tools was that, in both cases, a very small proportion of respondents believed (non-adopters)/claimed (adopters) that the use of CAT tools would decrease (non-adopters)/decreased (adopters) the prices of the translations undertaken by them, and that it was the main negative impact for both groups of respondents.

Table 8.14 **Impacts of adopting CAT tools (adopters and non-adopters)**

Impact on...	AD Rank	AD Score	NA Rank	NA Score
Quality of translations	1	89.5%	7	32.3%
Translator's productivity	2	88.9%	5	46.9%
Volume of work undertaken	3	72.2%	4	50%
Volume of work offered to translators by clients	4	66.7%	1	53.1%
Translator's turnover	5	61.1%	2	51.5%
Size of translator's customer base	6	42.1%	3	50%
Number of clients	7	42.1%	6	43.8%
Prices translators charge for their work	8	5.6%	8	3.1%

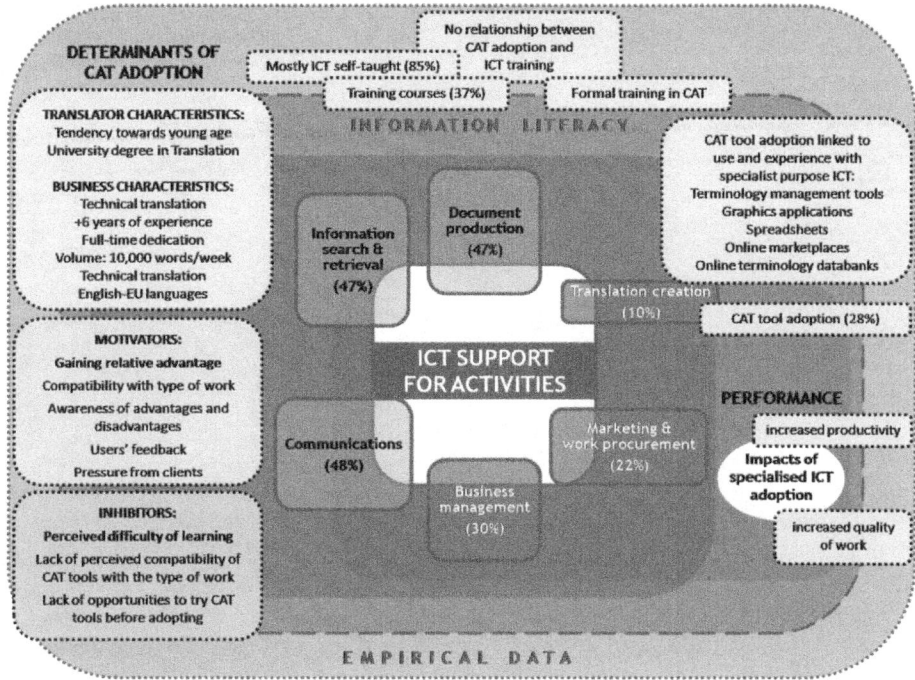

Figure 8.7 Research model enhanced by empirical data

These results reflected a higher degree of scepticism about the positive impacts of using CAT tools among those translators who had not adopted these tools. The analysis of the impacts among adopters and non-adopters, and the comparison of the results showed that the beliefs of non-adopters did not match with the impacts of CAT tool adoption reported by the adopters of these tools. Actually, the main impacts highlighted by adopters (increased productivity and quality of translations) were not considered as a potential impact by more than half of the non-adopters.

8.8 Summary of needs and perspectives

Figure 8.7 reflects a summarised view of the empirical findings reported in this chapter applied to the research model defined to investigate the adoption of ICT among MIPs and the issues surrounding the ICT sophistication of the translation businesses under study.

Part Three

Multilingual Information Management: matching needs and perspectives

"The best resources are surely those generated by professionals themselves during their professional activity, through arduous searches."
(Sales and Pinto 2011, p. 254)

In the previous parts of this book, a picture has been provided of how information intertwines with multilingual information professionals. First, by defining the ICT-based environment that surrounds multilingual information related workplaces, mostly freelance translators, the information literacy paradigm that surrounds their professional practice and a strategic approach that can help us to understand how and why ICT is adopted. Then, a research framework is presented that helps understand how to interact within this multilingual environment and that includes a number of research instruments based on empirical research. In this final part, a proposal is made of how to critically assess information and technology needs and how to strategically integrate them within a web-based environment in the form of a personal learning environment in the workplace, from an educational and lifelong learning perspective, eventually leading to the conception of an IS addressed to the particular setting of MIPs in the form of the proposed Multilingual Information Management System.

From PLEs to PLWEs: a Multilingual Information Management System

The proposal of a Multilingual Information Management System (MIMS from here on) introduced here is the result of the interdisciplinary approach to Multilingual information Management described throughout the book. It draws on the IS and IL concepts reviewed in the first part of the book, as well as in the study of the professional environment of MIPs, first addressed from the existing body of literature, also in the first part, and then enhanced by the development of a framework based on empirical research in the second part of the book. Therefore, the proposed MIMS enlarges the scope of core translation production within a multilingual working environment to add systems that support management and administrative activities as well as information and knowledge management within professional practice, and contributes to improving productivity, efficiency and decision-making.

Personalised information systems in knowledge-based working settings have been criticised for providing limited success because the systems have been designed according to what the technology can do rather than focusing on the user's perspective of what the work requires (Kuhlthau and Tama, 2001, p. 40). In line with the proposed approach so far and with a focus on the MIP and his or her business, rather than on the features of the available technology, the conceptualisation of the MIMS that follows takes into consideration the approach provided by Personal Learning Environments, their application to professional contexts, and the particular environment of MIM.

9.1 Personal Learning Environments (PLEs)

Information overload has been present along the discussion of previous chapters due to the problems it causes for those professionals working in multilingual communication. However, it also has the view that this larger-than-ever breadth of information sources and ICT within our reach provides a richer-than-ever environment for accessing information and knowledge and for facilitating continuous learning, i.e. lifelong learning, if the tools and resources – material and human – are used in a suitable way. This view aligns very much with that of Personal Learning Environments (PLEs), which Adell and Castañeda essentially defined as "the set of tools, sources of information, connections, and activities that each individual regularly uses to learn" (2010, author's translation).[1]

PLEs as such have always existed among people, although there has been a renewed academic interest in them since the advent of Web 2.0 and they have arisen as an educational approach to integrating the current environment of learners in the context of the information society, heavily influenced by the empowerment of social networks and the

---

[1] The original quotation reads "es el conjunto de herramientas, fuentes de información, conexiones y actividades que cada persona utiliza de forma asidua para aprender" (Adell and Castañeda 2010, p. 23).

ease of accessing information through Web 2.0 (Buchem et al., 2011). It is important to highlight that PLEs are not a technology or a system themselves, but an approach to taking advantage of the current ICT to teach and learn (Castañeda and Adell, 2013, p. 29). Another important assumption highlighted by Castañeda and Adell is that PLEs are not tightly subject to any particular didactic prescriptions, but rather on the contrary, they look to build each individual's learning environment by dynamically making knowledge explicit, managing it and generating connections with the components of this pedagogical ecosystem, be they sources of information or other individuals sharing knowledge. From a pedagogical theory perspective, the PLE approach is closely related to the ideas advocated by Vygotsky's socioconstructivism (1978) and with a more recent pedagogical current of connectivism (Siemens, 2005; Calvani, 2009; Downes, 2010). Thus, PLEs require a proactive approach to learning, both from teachers and learners, based on social interaction with peers, available sources of information, and ICT. In fact, learners can contribute to the environment with their input and eventually assume a teaching role, and similarly, teachers who are actively involved in a PLE will also become learners at some point during the teaching-learning process. This bi- or multi-directional flow of knowledge and learning is what makes PLEs a perfect candidate for facilitating lifelong learning, "learning to learn in the digital age" in Castañeda and Adell's words (2013, p. 22). Given this social commitment and involvement of the learning process within a technology-based environment an updated and more comprehensive definition has recently been provided by Attwell et al. (2013, p. iv): "a pedagogical approach with many implications for the learning processes, underpinned by a 'hard' technological base. Such a technopedagogical concept can benefit from the affordances of technologies, as well as from the emergent social dynamics of new pedagogic scenarios."

To understand how a PLE works it is important to be aware of the three interrelated pillars that integrate this ecosystem (see Figure 9.1.), namely, accessing information and reading it, reflecting and writing about it, and sharing knowledge and learning.

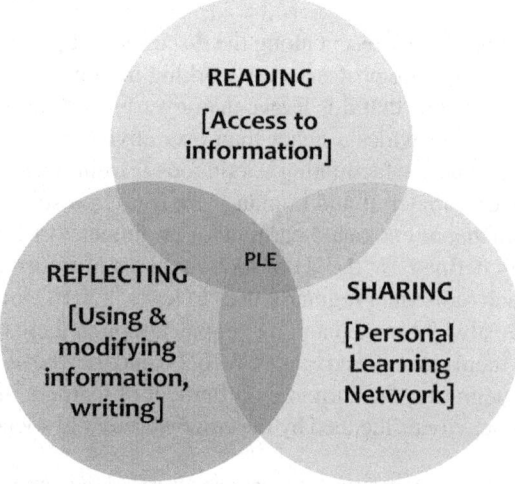

Figure 9.1 Conception of PLE
(adapted from Castañeda and Adell, 2013, author's translation)

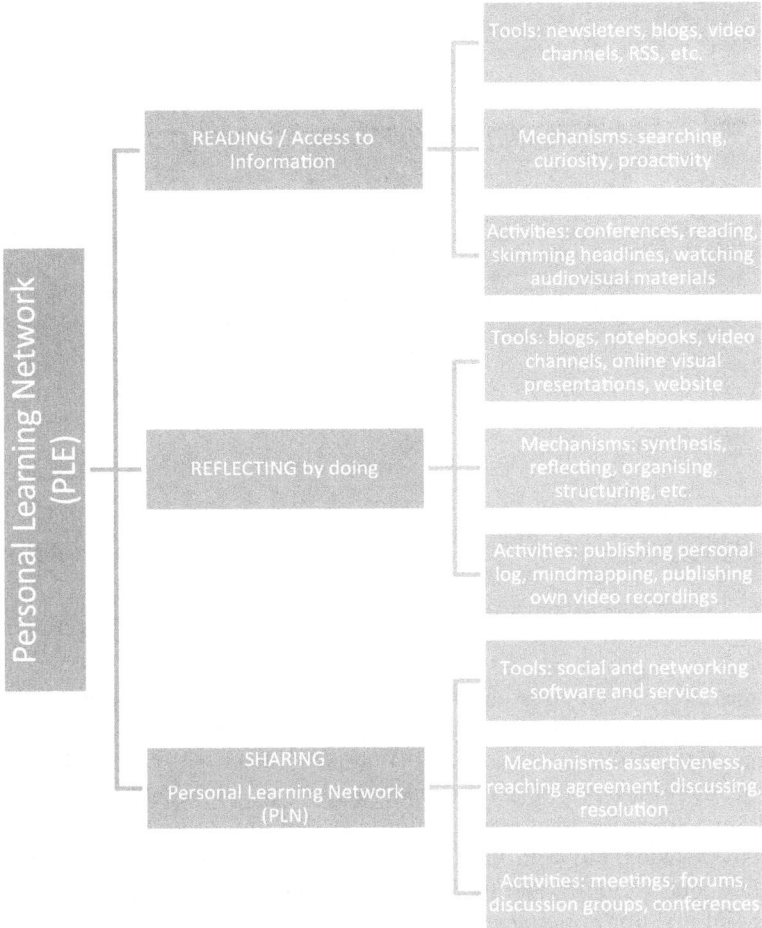

Figure 9.2 Components of a PLE
(adapted from Castañeda and Adell, 2013, author's translation)

In giving an example of what a PLE involves, a myriad of diagrams and visualisations of individual PLEs can be found on the Internet;[2] however, from a formal and rigorous analysis a comprehensive conceptualisation is presented by Castañeda and Adell (2013, p. 20), which also gives an example and has evolved from their work in 2010 to the version in 2013. Figure 9.2 presents an adaptation of their conceptualisation of a PLE.

A practice that is becoming more and more popular among online communities as a way to share information and that falls into the kind of practices that are part of a PLE is information curation (c.f. Chapter 6, section 6.3.1.4.). Originating in the context of the modern version of the old practice of archiving (Joint Information Systems

---

[2] A simple search on Google Images using the keyword "ple" provides a very large amount of representations from the more than 14 million results.

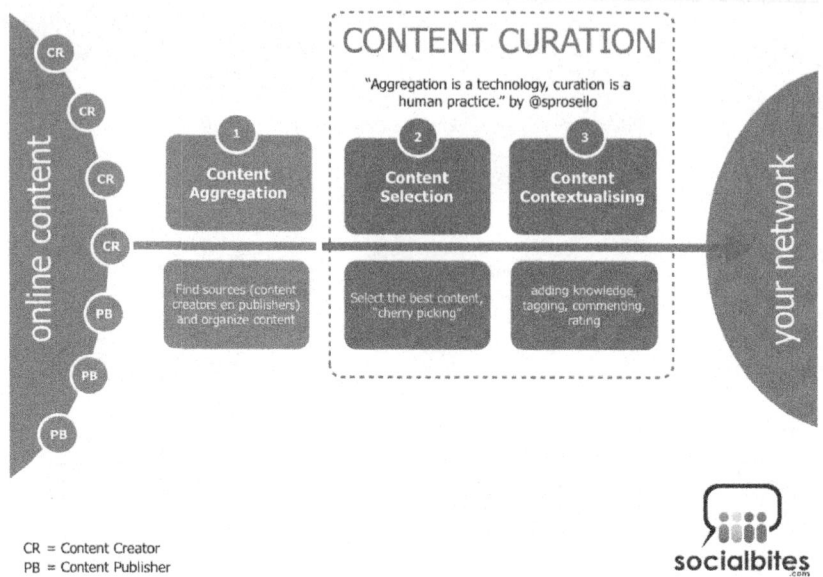

Figure 9.3 Content curation process

Committee, 2003), curation is understood in the sense of preserving personal digital archives for future consumption (text- or multimedia-based), as opposed to the idea of immediate consumption as we access it (Whittaker, 2011, p. 5). Actually, Whittaker states that for most types of personal information, people's behaviour seems to be much closer to curation than consumption. Similarly to PLEs, information archives have always existed, but with the growth of digital content, the ease for accessing information and storing it, and the proliferation and empowerment of online social networks, information curation is becoming a growing trend which is actually affecting the way people consume goods (Colombani and Videlaine, 2013). Figure 9.3. shows a diagram about content curation published by Socialbites.com (website not available anymore) that presents a process very similar to that of PLEs.

9.2 Personal Learning and Working Environments (PLWEs)

PLEs, if viewed from a systemic point of view (c.f. Chapter 2, section 2.1.3.), share a number of similarities with information systems. A PLE is a system from the very moment it is an environment that connects several components to work together towards the objective of learning, there are continuous inputs and outputs – a multidirectional flow of information. It is also closely related to the knowledge generation processes of information systems, since the main asset being exchanged is information that, eventually, generates knowledge. There is also an element in common with IS strategy, since the way a PLE is created and the way it changes and evolves has a lot to do with the decisions being made at its conception, e.g. how a PLE is organised, and with the goals that are set initially and at a longer term, i.e. there is a main learning objective for developing a PLE and there are choices that are made to best accomplish it. Therefore, there is strategic thinking guiding the conception of a PLE (c.f. first conception of IS strategy in Chapter 5, section 5.2.), there is a tactical approach to decide what assets should be part of a PLE (c.f. second conception), i.e. what components form the PLE in terms of tools, resources, people, and there is an overall perspective of a PLE that socially defines and situates each individual behind it (c.f. third conception).

PLEs have been conceived to address the educational scenario of our current society, where Web 2.0 and the social face of the Internet have transformed the way in which people are connected to each other and share information and knowledge. This fact has not only affected the way we learn, but also the way we work; therefore, PLE-like approaches cannot only be useful for self-development and learning, but also for continuing this learning throughout ones' professional activity and for supporting information management as part of professional practice (Ardichvili et al., 2003; Jarche 2008). A previous conception of this view from a Knowledge Management angle was the figure of an eProfessional, as a variation of the eLearner, and tagged this application of the PLE concept to a working setting as a Personal Learning and Work Environment (PLWE) (Rubio et al., 2011).

Scholars have also recently pointed towards adopting a PLE-like approach in higher education training as an early contact to a setting which is similar to the professional practice context of translators (Cánovas 2013). This view, in addition to replicating part of the social environment of translators, also shows an information literacy approach towards creating a personal knowledge base which arises from the learning experience acquired through formal training and that should grow and evolve in parallel to the professional behind it. In this application of a PLE and an IL to professional practice further support is also required to make informed decisions about how strategically to mature this environment to meet business goals and provide competitive services, how to tactically operate within the working environment to be efficient at managing all the information and information sources, and how to convey this vision to others. This additional support can be provided in the form of an MIMS.

9.3 A Multilingual Information Management System

Returning to the MIM context, the application of the PLWE approach with the addition of the IS and IL input defines the holistic understanding of an MIMS that can enhance the professional practice of MIPs, as an intense knowledge activity (Sales et al. forthcoming) and that does not only rely on material sources of information, but also on human ones (Montalt i Resurrecció, 2009, p. 174). This is particularly relevant and required for those working in specialised contexts, such as legal, medical or technical multilingual communication (García Izquierdo and Borja Albi, 2009; Borja, 2005).

As stated at the beginning of the book, both formal and informal communication can lead to the generation of knowledge (c.f. 2.1.2), although if information is not formalised, it is much more difficult to manage it so it can be accessed when is required or can support decision-making. As stated by Li (2009, p. 19), "In today's information society, heterogeneous data and information can be dynamically accessed, converted, disseminated, distributed, located, processed, and stored across diverse applications, channels, databases, networks, platforms and systems," but to do it efficiently within a working setting it needs to be suitably managed. This is also the case in MIM, where information comes from a vast range of sources that needs to be organised to be processed and eventually retrieved.

As it happens with the competition between finding MT systems that replace human translators (c.f. efforts towards FAHQT in Chapter 3) and developing tools that support translators' work at their core function, i.e. CAT tools (c.f. Chapter 3), in the IS field, Decision Support Systems were designed to support rather than replace decisions within an organisation. Similar to this overall resemblance, DSSs involve flexible interactive access to data. They are designed with an understanding of the requirements of the decision makers and the decision-making process in mind to deliver the right information at all levels of operation, including end-users.

Information access has greatly improved due to technological developments; however, an integrated system such as the proposed MIMS should aim at improving MIP effectiveness and efficacy by providing support at the functions integrated within a MIM environment (c.f. Chapter 6, section 6.3.1. for a more detailed description of the ICT support for the activities behind the functions), particularly at the following strategy levels:

Tactical-operational level: administrative tasks that can be automated to some extent at all levels of operation are required in order to improve productivity and be able to focus on core functions. This involves, for example, from basic clerical and communication tasks to more advanced online communication activities, budgeting, invoicing, and tax and accounts management tasks (c.f. Chapter 6.3.1.4. and 6.3.1.6).

Tactical-strategic level of information management: all the information-related tasks that require a query to cover a need depend upon a reliable and effective system that captures, stores, organises and delivers the most suitable piece of information for each case, be it a business-related need or a language specific one. The former include, for example, client databases, finance reports, direct or social communications, and marketing materials. Among the latter knowledge bases, terminology collections, translation memories, multilingual corpora, reference materials, online communities or domain experts, or the generation of multilingual documents (c.f. Chapter 6.3.1.2. and 6.3.1.3.).

Managerial-strategic level: an increased support for tasks performed by MIP in the decision-making area is required for making timely and knowledgeable decisions at all levels of operation (c.f. Chapter 5.2.), from evaluating the value and feasibility of undertaking potential projects, to defining language-, culture- or market-specific strategies that will guide the whole linguistic transfer of information, and through the preparation of support resources to optimise the formulation and quality assessment of translations (c.f. Chapter 6.3.1.1., 6.3.1.3. and 6.3.1.5.).

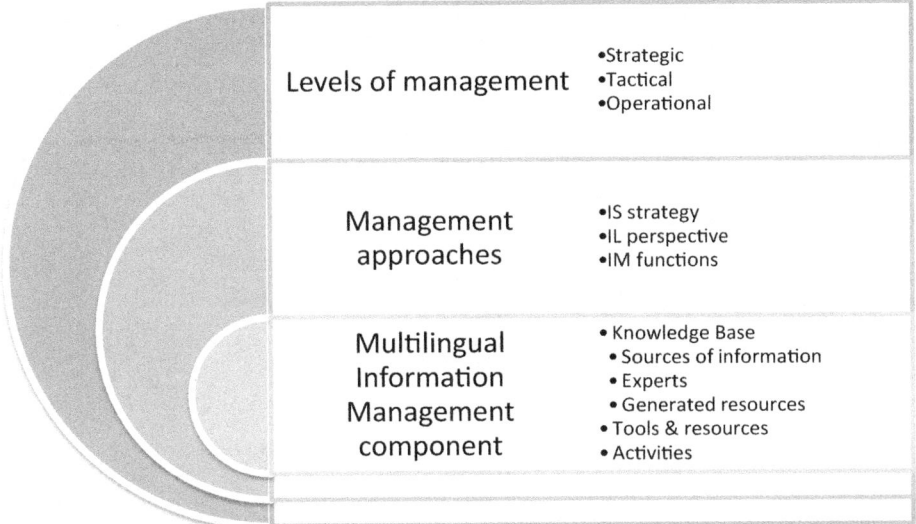

Figure 9.4 Components of an MIMS

9.4 Structure of an MIMS

In order to cover the needs of the system, a suitable structure should include the following components that have been presented in detail throughout the book (Figure 9.4):

- the levels of management required: strategic, tactical and operational;
- the approaches/views that regulate each of these levels: the information systems strategy, the information literacy approach and the information management activities;
- the core MIM component that includes an integrated view of the sources of information, tools, resources and people that are part of the business activity.

References

Abaitua, J. (1999). Quince años de traducción automática en España. *Perspectives: Studies in Translatology*, 7(2), 221–230.
ACRL (2000). Information Literacy Competency Standards for Higher Education. Available at: *http://www.ala.org/acrl/standards/informationliteracycompetency* [Accessed February 4, 2013].
Adell, J., & Castañeda, L. (2010). Los Entornos Personales de Aprendizaje (PLEs): una nueva manera de entender el aprendizaje. In R. Roig Vila & M. Fiorucci (Eds.), *Claves para la investigación en innovación y calidad educativas. La integración de las Tecnologías de la Información y la Comunicación y la Interculturalidad en las aulas. Strumenti di ricerca per l'innovazione e la qualità in ámbito educativo. La Tecnologie dell'informazione e della Comunicazione e l'interculturalità nella scuola*. Marfil – Roma TRE Universita degli studi. Available at: *http://digitum.um.es/jspui/handle/10201/17247* [Accessed April 23, 2014].
Aikat, D. "Deb", & Remund, D. (2012). Of Time Magazine, 24/7 media, and data deluge: The evolution of information overload theories and concepts. In J. B. Strother, J.M. Ulijn, & Z. Fazal (Eds.), *Information overload*. Hoboken, NJ, USA: John Wiley & Sons, Inc. (pp. 13–40). Available at: *http://doi.wiley.com/10.1002/9781118360491.ch2* [Accessed August 13, 2013].
Aikawa, T., et al. (2007). Impact of controlled language on translation quality and post-editing in a statistical machine translation environment. In *Proceedings of MT Summit XI* (pp. 1–7). Available at: *http://mt-archive.info/MTS-2007-Aikawa.pdf* [Accessed August 18, 2013].
Alcina, A., Soler, V., & Granell, J. (2007). Translation technology skills acquisition. *Perspectives: Studies in Translatology*, 15(4), 230–244.
ALPAC (1966). *Language and machines: computers in translation and linguistics*, Automatic Language Processing Advisory Committee, Division of Behavioral Sciences. National Research Council.
American Library Association Presidential Committee on Information Literacy (1989). Final report, available at: http://www.ala.org/acrl/publications/whitepapers/presidential [Accessed January 20, 2014].
Amstrong, J. S., & Overton, T. A. (1977). Estimating nonresponse bias in mail surveys. *Journal of Marketing Research*, 14, 396–402.
Andretta, S. (2005). *Information literacy: a practitioner's guide*. Oxford: Chandos Publishing.
ANECA (2004). *Libro blanco del título de grado en traducción e interpretación*, Agencia Nacional de Evaluación de la Calidad y Acreditación Madrid. Available at: www.aneca.es/media/150288/libroblanco_traduc_def.pdf.
Ardichvili, A., Cardozo, R., & Ray, S. (2003). A theory of entrepreneurial opportunity identification and development. *Journal of Business Venturing*, 18(1), 105–123.
Attwell, G., Castañeda, L., & Buchem, I. (2013). Guest editorial preface: special issue from the personal learning environments 2011 Conference. *International Journal of Virtual and Personal Learning Environments (IJVPLE)*, 4(4), iv–vii.
Austermühl, F. (2001). *Electronic tools for translators*. Manchester: St. Jerome Publishing.
Baker, W. H. (1987). Status of information management in small businesses. *Journal of Systems Management*, 38(4), 10–15.
Basili, C. (2008). Information and education policies in Europe: Key factors influencing information literacy academic policies in Europe. In C. Basili (Ed.), *Information literacy at the*

crossroad of education and information policies in Europe (pp. 18–32). Roma: Consiglio nazionale delle ricerche. Istituto di ricerca sull'impresa e lo sviluppo.

Bawden, D. (2001). Information and digital literacies: a review of concepts. *Journal of Documentation*, 57(2), 218–259.

Benbasat, I., et al. (1984). A critique of the stage hypothesis: theory and empirical evidence. *Communications of the ACM*, 27(5), 476–485.

Benbasat, I., Dexter, A. S., & Mantha, R. W. (1980). Impact on organizational maturity on information systems skills needs. *MIS Quarterly*, 4(1), 21–34.

Beynon-Davies, P. (2002). *Information systems: an introduction to informatics in organisations*. Palgrave Macmillan Limited.

Bhatt, G. D., & Grover, V. (2005). Types of information technology capabilities and their role in competitive advantage: an empirical study. *Journal of Management Information Systems*, 22(2), 253–277.

Blatt, A. (1998). Workflow using linguistic technology at the translation service of the European Commission. In *Workshop of the European Association for Machine Translation* (pp. 7–18). WHO.

Bocij, P., et al. (1999). *Business information systems. technology, development and management*. London: Financial Times Management.

Borgman, C. L. (2007). *Scholarship in the digital age*. MA: MIT Press.

Borja, A. (2005). Organización del conocimiento para la traducción jurídica a través de sistemas expertos basados en el concepto de género textual. In I. García Izquierdo (Ed.), *El género textual y la traducción. Reflexiones teóricas y aplicaciones pedagógicas* (pp. 37–69). Berna: Peter Lang.

Bourner, T., et al. (1983). The diffusion of microelectronic technology in south-east england. In D. L. Bosworth (Ed.), *The employment consequences of technological change* (pp. 97–109). London: Macmillan.

Bowker, L. (2010). Can Machine Translation meet the needs of official language minority communities in Canada? A recipient evaluation. *Evaluation of Translation Technology*, 8, 123.

Bowker, L. (2002). *Computer-aided translation technology: a practical introduction*. Ottawa: University of Ottawa Press.

Brace, C. (2000). Language automation at the European Commission. In Sprung R.C. (Ed.), *Translating into success: cutting-edge strategies for going multilingual in a global age. Amsterdam* (pp. 219–224). Philadelphia: John Benjamins.

Brace, C., Vasconcellos, M., & Miller, L. C. (1995). MT users and usage: Europe and the Americas. *MT News International*, 12, 14–19.

Brannen, J. (1995). *Mixing methods: qualitative and quantitative research*. Aldershot: Avebury.

Bruce, C. S. (1997). *The seven faces of information literacy*. Adelaide: Auslib Press.

Bruce, C. S. (1999). Workplace experiences of information literacy. *International Journal of Information Management*, 19(1), 33–47.

Bruce, C. S., Candy, P. C., & Klaus, H. (2000). *Information literacy around the world: advances in programs and research*, Wagga Wagga, N.S.W.: Centre for Information Studies, Charles Sturt University.

Bruns, A., & Burgess, J. E. (2011). The use of Twitter hashtags in the formation of ad hoc publics. In: *ARC Centre of Excellence for Creative Industries and Innovation; Creative Industries Faculty; Institute for Creative Industries and Innovation. 6th European Consortium for Political Research General Conference*. University of Iceland, Reykjavik. Available at: http://www.ecprnet.eu/conferences/general_conference/reykjavik/ [Accessed August 21, 2013].

Bryman, A. (1989). *Research methods and organization studies*. London: Routledge.
Bryman, A., & Bell, E. (2003). *Business research methods*. Oxford: Oxford University Press.
Buchem, I., Attwell, G., & Torres, R. (2011). Understanding Personal Learning Environments: literature review and synthesis through the Activity Theory lens. In: *The PLE Conference 2011*. Southampton, UK, pp. 1–33. Available at: *http://journal.webscience.org/658/* [Accessed April 23, 2014].
Buckland, M. K. (1991). Information as thing. *Journal of the American Society for Information Science*, *42*(5), 351–360.
Bundy, A. (Ed.) (2004). *Australian and New Zealand Information Literacy Framework: principles, standards and practice* (2nd ed.). Adelaide: Australian and New Zealand Institute for Information Literacy. Available at: *http://archive.caul.edu.au/info-literacy/InfoLiteracy-Framework.pdf* [Accessed February 4, 2013].
Buneman, P., et al. (2008). Curated databases. In Proceedings of the twenty-seventh ACM SIGMOD-SIGACT-SIGART symposium on principles of database systems. PODS '08. New York, NY, USA: ACM, pp. 1–12. Available at: *http://doi.acm.org/10.1145/1376916.1376918* [Accessed August 22, 2013].
Burgess, R. G. (1984). *In the field: An introduction to field research*. London: George Allen and Unwin.
Buyya, R., Broberg, J., & Goscinski, A. M. (2010). *Cloud computing: principles and paradigms*. New York, NY, USA: John Wiley & Sons.
Byrd, T. A., & Turner, D. E. (2001). An exploratory examination of the relationship between flexible IT infrastructure and competitive advantage. *Information & Management*, *39*(1), 41–52.
Caldeira, M. M., & Ward, J. M. (2002). Understanding the successful adoption and use of IT/IS in SMEs: an explanation from Portuguese manufacturing industries. *Information Systems Journal*, *12*(2), 121–152.
Calvani, A. (2009). Connectivism: new paradigm or fascinating pot-pourri? *Journal of E-learning and Knowledge Society*, *4*(1). Available at: *http://www.editlib.org/p/43293/article_43293.pdf* [Accessed April 23, 2014].
Cánovas, M. (2013). Los entornos personales de aprendizaje (PLE) en la formación de traductores: pedagogía y tecnología. *Tradumàtica: traducció i tecnologies de la informació i la comunicació*, *0*(11), 257–266.
Case, D. O. (2012). *Looking for information: a survey of research on information seeking, needs and behavior*. Emerald Group Publishing.
Castañeda, L., & Adell, J. (2013). La anatomía de los PLEs. In L. Castañeda & J. Adell (Eds.) *Entornos Personales de Aprendizaje: claves para el ecosistema educativo en red*. Alcoy: Editorial Marfil, pp. 11–28.
Catts, R., & Lau, J. (2008). *Towards information literacy indicators*. Paris: UNESCO.
Champollion, Y. (2003). Convergence in CAT: blending MT, TM. OCR & SR to boost productivity. In *Proceedings of the International Conference Translating and the Computer 25*, 20-21 November 2003, London. London: Aslib.
Chan, Y. E., et al. (1997). Business strategic orientation, information systems strategic orientation, and strategic alignment. *Information Systems Research*, *8*(2), 125–150.
Chan, Y. E., & Reich, B. H. (2007). IT alignment: what have we learned? *Journal of Information technology*, *22*(4), 297–315.
Chanod, J. (1998). Multilingual tools at the Xerox Research Centre. In *1998 workshop of the European Association for Machine Translation* (pp. 73–84). Geneva: WHO.
Chen, D. Q., et al. (2010). Information systems strategy: reconceptualization, measurement, and implications. *MIS Q*, *34*(2), 233–259.

Cheney, P. H., & Dickson, G. W. (1982). Organizational characteristics and information systems: an exploratory investigation. *Academy of Management Journal*, 25(1), 170–184.

Chenhall, R. H., & Morris, D. (1986). The impact of structure, environment, and interdependence on the perceived usefulness of management accounting systems. *Accounting Review*, 61(1), 16–35.

Cheuk, W. B. (1998). Exploring information literacy in the workplace: a qualitative study of engineers using the sense-making approach. *International forum on information and documentation*, 23(2), 30–38.

Cheuk, B. (2002). *Information literacy in the workplace context: issues, best practices and challenges. White Paper prepared for UNESCO, the U.S. National Commission on Libraries and Information Science, and the National Forum on Information Literacy*, for use at the Information Literacy Meeting of Experts, Prague, The Czech Republic. Available at: http://www.nclis.gov/libinter/infolitconf&meet/papers/cheuk-fullpaper.pdf.

Churchill, G. A. (1999). *Marketing research: methodological foundations* (7th ed.). London: Harcourt College Publishers.

Churchill, N. C., Kempster, J. H., & Uretsky, M. (1969). *Computer based information systems for management: a survey*. New York: National Association of Accountants.

Clarke, P. (2000). The Internet as a medium for qualitative research. *South African Journal of Information Management*, 2(2/3), .

Collins English Dictionary. (2014). Available at: http://www.collinsdictionary.com/dictionary/english/information.

Colombani, L., & Videlaine, F. (2013). *The age of curation: From abundance to discovery*, Bain & Company. Available at: http://www.bain.com/publications/articles/the-age-of-curation-from-abundance-to-discovery.aspx [Accessed April 23, 2014].

Cooley, P. L., Walz, D. T., & Walz, D. B. (1987). A research agenda for computers and small business. *American Journal of Small Business*, 11(3), 31–42.

Cornellá, A. (1999). A mayor desarrollo informacional, menor infoxicación. *El Profesional de la Información*, 8(9), 42–44.

Cornford, T., & Smithson, S. (1996). Project research in information systems. *A student's guide*. London: MacMillan Press.

Cragg, P. B. (1990). *Information technology and small firm performance*. Ph. D. thesis. Loughborough: Loughborough University of Technology.

Cragg, P. B., & King, M. (1993). Small-firm computing: motivators and inhibitors. *MIS Quarterly*, 17(1), 47–59.

Cragg, P. B., & Zinatelli, N. (1995). The evolution of information systems in small firms. *Information & Management*, 29, 1–8.

Crossan, F. (2003). Research philosophy: towards an understanding. *Nurse Researcher*, 11(1), 46–55.

Daniel, E., Wilson, H., & Myers, A. (2002). Adoption of e-commerce by SMEs in the UK: towards a stage model. *International Small Business Journal*, 20(3), 253–270.

Davis, F.D. (1986). *A technology acceptance model for empirically testing new end user information systems: theory and results* PhD Thesis. Massachusetts: Massachusetts Institute of Technology, Sloan School of Management .

Davis, G.B. (2000). Information systems conceptual foundations: looking backward and forward. In R. Baskerville, J. Stage & J. DeGross (Eds.) *Organizational and social perspectives on information technology*. Springer, pp. 61–82.

DeLone, W. H. (1988). Determinants of success for computer usage in small business. *MIS Quarterly*, 12(1), 51–61.

DeLone, W. H. (1981). Firm size and the characteristics of computer use. *MIS Quarterly*, 5(4), 65–77.

Denzin, N. (1970). *The research act in sociology*. London: Butterworth.
Department of Translation and Communication at the Universitat Jaume I, (2013). Master's Degree in Medical and Healthcare Translation. Available at: *http://www.uji.es/UK/infoest/estudis/postgrau/oficial/tmedsan.html* [Accessed February 17, 2014].
Díaz Fouces, O., & García González, M. (2008). *Traducir (con) software libre*. Granada: Comares.
Dillman, D. A. (2000). *Mail and internet surveys: the tailored design method*. NY: John Wiley & Sons.
Dillman, D. A. (1978). *Mail and telephone surveys: the total design method*. NY: John Wiley & Sons.
Dillon, S., & Fraser, J. (2006). Translators and TM: an investigation of translators' perceptions of translation memory adoption. *Machine Translation, 20*(2), 67–79.
Downes, S. (2010). *Learning networks and connective knowledge*. Available at: http://philpapers.org/rec/DOWLNA [Accessed April 23, 2014].
Drago, I., et al. (2012). Inside dropbox: understanding personal cloud storage services. In *Proceedings of the 2012 ACM conference on Internet measurement conference*. IMC '12. New York, NY, USA: ACM, pp. 481–94. Available at: *http://doi.acm.org/10.1145/2398776.2398827* [Accessed August 21, 2013].
Drucker, P. F. (1993). *Post-capitalist society*. New York, NY: HarperBusiness.
Duncan, N. (1995). Capturing IT infrastructure flexibility: A study of resource characteristics and their measure. *Journal of Management Information Systems, 12*(2), 37–57.
Dunne, K. J. (2006). *Perspectives on localization*. Amsterdam, Philadelphia: John Benjamins Publishing.
EAGLES (1996). Evaluation of natural language processing systems. Final Report. EAGLES Document EAG-EWG-PR2.
Earl, M. J. (1989). *Management strategies for information technology*. London: Prentice-Hall.
Easterby-Smith, M., Thorpe, R., & Lowe, A. (1999). *Management research. an introduction*. London: SAGE Publications.
Easton, G., et al. (1982). *Small computers in small companies*. Lancaster: Department of Marketing, University of Lancaster.
Ein-Dor, P., & Segev, E. (1982). Organizational context and MIS structure: some empirical eEvidence. *MIS Quarterly, 6*(3), 55–68.
Eisenberg, M., & Berkowitz, R. E. (1990). *Information problem-solving: the Big Six Skills approach to library & information skills instruction*. Norwood, NJ: Ablex Pub. Corp.
Elmborg, J. (2006). Critical information literacy: implications for instructional practice. *The Journal of Academic Librarianship, 32*(2), 192–199.
EMT expert group (2009). *Competences for professional translators, experts in multilingual and multimedia communication,* Brussels. Available at: http://ec.europa.eu/dgs/translation/programmes/emt/key_documents/emt_competences_translators_en.pdf [Accessed February 17, 2014].
Enriquez Raido, V. (2011). Developing web searching skills in translator training. *Revista Electrónica de Didáctica de la Traducción y la Interpretación, (6)*, 57–77.
Esselink, B. (2000). *A practical guide to localization Rev. ed.* Amsterdam, Philadelphia: John Benjamins.
Esselink, B. (2003). Localisation and translation. In H. Somers (Ed.), *Computers and translation: a translator's guide* (pp. 67–86). Amsterdam, Philadelphia: John Benjamins.
European Commission, Directorate-General for Education and Culture, 2003. *Implementation of "Education and Training 2010" work programme: progress report of the working group "Basic Skills, Entrepreneurship and Foreign Languages,"* Brussels: European Commission

Farhoomand, F., & Hrycyk, G. P. (1985). The feasibility of computers in the small business environment. *American Journal of Small Business, 9*(4), 15–22.

Faught, K. S., Whitten, D., & Green, K. W. (2004). Doing survey research on the Internet: yes, timing does matter. *Journal of Computer Information Systems, 44*(3), 26–34.

Fenner, A. (2000). The choices facing translators. *Institute of Translation and Interpreting Bulletin, April 2000,* 9.

Fiederer, R., & O'Brien, S. (2009). Quality and machine translation: a realistic objective. *The Journal of Specialised Translation, 11,* 52–74.

Folaron, D. (2006). A discipline coming of age in the digital age. In K. J. Dunne (Ed.). *Perspectives on localization.* American Translators Association Scholarly Monograph Series XIII (pp. 195–219). Amsterdam, Philadelphia: John Benjamins.

Frankfort-Nachmias, C., & Nachmias, D. (1992). *Research methods in the social sciences* (4th ed.). Kent: Edward Arnold.

Fraser, J. (2001). Rolls Royce quality at Lada prices: a survey of the working conditions and status of freelance translators. In L. Desblache (Ed.), *Aspects of specialised translation* (pp. 31–47). La Maison du dictionnaire.

Fraser, J., & Gold, M. (2001). 'Portfolio workers': autonomy and control amongst freelance translators. *Work, Employment & Society, 15*(4), 679–697.

Fraser, J., & Gold, M. (2000). Rainy Sundays and sunny Tuesdays: freelance translators' views on their employment status. *Institute of Translation and Interpreting Bulletin, April 2000,* 2–8.

De la Fuente, R. (2012). Posedición, ¿cambio de paradigma? *Tradumàtica: traducció i tecnologies de la informació i la comunicació*(10), 147–149.

Fulford, H. (2001). Translation tools: an exploratory study of their adoption by UK freelance translators. *Machine Translation, 16*(4), 219–232.

Fulford, H. (2002a). Freelance translators and machine translation: an investigation of perceptions, uptake, experience and training needs. In *6th European Association of Machine Translation Workshop.* UMIST, pp. 117–22.

Fulford, H. (2002b). *Terminology resources on the world wide web.* Paper presented at the Translating and Interpreting Fair, Institute of Linguists, September 2002. University of Westminster, London.

Fulford, H., & Granell-Zafra, J. (2003). Internet skills and translation: training freelance translators to explore, exploit and evaluate the potential of web-based resources. In S. Posteguillo et al. (Eds.), *Internet in linguistics, translation and literary studies* (pp. 223–239). Castellón de la Plana: Publicacions de la Universitat Jaume I.

Fulford, H., & Granell-Zafra, J. (2004). The freelance translator's workstation: an empirical investigation. In *Broadening horizons of machine translation and its applications* (pp. 53–61). Malta: Foundation for International Studies.

Fulford, H., & Granell-Zafra, J. (2005). Translation and technology: a study of UK freelance translators. *JoSTrans, The Journal of Specialised Translation*(4), 2–17.

Fulford, H., & Granell-Zafra, J. (2008). The Internet in the freelance translator's workflow. *International Journal of Translation (IJT), 20*(1–2 July–Dec issue), 5–18.

Fulford, H., Höge, M., & Ahmad, K. (1990). *User requirements study. Final report for Workpackage 3.3,* EC ESPRIT II programme for project No. 2315. (Translator's Workbench Project).

Gable, G. G. (1991). Consultant engagement for first time computerization: a proactive client role in small businesses. *Information & Management, 20,* 83–93.

Gaiser, T. J. (1997). Conducting on-line focus groups. *Social Science Computer Review, 15*(2), 135–144.

Galliers, R. D. (1992). Choosing information systems research approaches. In R. D. Galliers (Ed.) *Information systems research - issues, methods and practical guidelines.* UK: Blackwell Scientific Publications.

Galliers, R. D. (1991). Strategic information systems planning: myths, reality and guidelines for successful implementation. *European Journal of Information Systems*, *1*(1), 55–64.

Galliers, R. D., & Land, F. F. (1987). Choosing appropriate information systems research methodologies. *Communications of the ACM*, *30*(11), 901–902.

Garcia, I. (2009). Beyond translation memory: computers and the professional translator. *The Journal of Specialised Translation*, *12*(12), 199–214.

Garcia, I. (2005). Long term memories: Trados and TM turn 20. *The Journal of Specialised Translation*, *4*, 18–31.

Garcia, I. (2011). Translating by post-editing: is it the way forward? *Machine Translation*, *25*(3), 217–237.

García Izquierdo, I., & Borja Albi, A. (2009). La gestión de la documentación multilingüe en entornos profesionales. *Lynx: Panorámica de estudios lingüísticos*, *8*, 3–28.

Gaspari, F. (2004). Online MT services and real users' needs: an empirical usability evaluation. In *Machine Translation: From Real Users to Research*. Springer, pp. 74–85. Available at: *http://link.springer.com/chapter/10.1007/978-3-540-30194-3_9* [Accessed February 26, 2014].

Gaspari, F., & Hutchins, J. (2007). Online and free! Ten years of online machine translation: origins, developments, current use and future prospects. *Proceedings of the Machine Translation Summit XI*, 199–206.

Gliner, J. A., & Morgan, G. A. (2000). *Research methods in applied settings*. Mahwah, NJ: Lawrence Erlbaum Associates.

Goad, T. W. (2002). *Information literacy and workplace performance*. Greenwood Publishing Group.

Gómez-Hernández, J.A. (2010). University libraries and the development of lecturers' and students' information competencies. *RUSC. Universities and Knowledge Society Journal*, *7*(2). Available at: *http://rusc.uoc.edu/ojs/index.php/rusc/article/view/v7n2-gomez* [Accessed April 15, 2014].

Gonzalo García, C. (2004). Fuentes de información en línea para la traducción especializada. In C. Gonzalo García, & V. García Yebra (Eds.), *Manual de documentación y terminología para la traducción especializada* (pp. 275–307). Madrid: Arco Libros.

Göpferich, S. (2009). Towards a model of translation competence and its acquisition: the longitudinal study TransComp. In S. Göpferich, A. L. Jakobsen, & I. M. Mees (Eds.), *Behind the mind: methods, models and results in translation process research* (pp. 11–38). Samfundslitteratur.

Granell-Zafra, J. (2006). *The adoption of computer-aided translation tools by freelance translators in the UK*. Ph. D. Thesis. Loughborough: Loughborough University.

Gremillion, L. (1984). Organizational Size and Information System Use: An Empirical Study. *Journal of Management Information Systems*, *1*(2), 4–17.

Gul, F. A. (1991). The effects of management accounting systems and environmental uncertainty on small business managers' performance. *Accounting and Business Research*, *22*(85), 57–61.

Hair, J. F., et al. (1998). *Multivariate data analysis* (5th ed.). Upper Saddle River, N.J: Prentice Hall.

Hamilton, S., & Chervany, N. L. (1981). Evaluating information system effectiveness part I: comparing evaluation approaches. *MIS Quarterly*, *5*(3), 55–69.

Hanna, R., Rohm, A., & Crittenden, V. L. (2011). We're all connected: the power of the social media ecosystem. *Business Horizons*, *54*(3), 265–273.

Harper, D. (2014). communication. *Online Etymology Dictionary*. Available at: *http://www.etymonline.com* [Accessed January 27, 2014].

Hatim, B., & Mason, I. (1990). *Discourse and the translator*. New York: Longman.

Henderson, J. C., & Venkatraman, N. (1993). Strategic alignment: leveraging information technology for transforming organizations. *IBM Systems Journal*, *32*(1), 4–16.

Hepworth, M., & Smith, M. (2008). Workplace information literacy for administrative staff in HE. *Australian Library Journal*, 57 (3), pp. 212-236.
Heyn, M. (1996). Integrating machine translation into translation memory systems. In *Proceedings of the EAMT Machine Translation Workshop*, TKE'96. pp. 113–26. Available at: http://www.mt-archive.info/EAMT-1996-WS.pdf#page=113 [Accessed February 26, 2014].
Heyn, M. (1998). Translation memories: insights and prospects. In L. Bowker et al. (Eds.), *Unity in diversity? Current trends in translation studies* (pp. 123–136). Manchester: St. Jerome Publishing.
Höcker, M. (2003). *eColore translation memory survey 2003*, Berlin: Bundesverband der Dolmetscher und Übersetzer (BDÜ). Available at: http://ecolore.leeds.ac.uk/downloads/2003.05_bdue_survey_analysis.doc.
Höge, M. (2002). *Towards a framework for the evaluation of translators aids systems*. Ph. D. thesis. Helsinki: Helsinki University.
Holland, C. P. et al. (2004). Marketing translation services internationally: exploiting IT to achieve a smart network. *Journal of Information Technology*, 19(4), 254–260.
Holmes, J. S. (1988). *Translated! - Papers on literary translation and translation studies*. Amsterdam: Rodopi.
Howe, J. (2006). The rise of crowdsourcing. *Wired*, 14(6). Available at: http://www.wired.com/wired/archive/14.06/crowds.html [Accessed August 7, 2013].
Hunt, T. (2002). Translation support software: the next generation replacement to CAT tools. *ATA Chronicle*, 31(1), 49–52.
Hunt, T. (2003). Translation technology failures and future. *The LISA Newsletter: Globalization Insider, XI* (4.6).
Hurtado Albir, A. (2007). Competence-based curriculum design for training translators. *The Interpreter and Translator Trainer*, 1(2), 163–195.
Hussin, H. (1998). *Alignment of business strategy and IT strategy in small businesses*, Ph. D. Thesis. Loughborough: Loughborough University.
Hussin, H., King, M., & Cragg, P. B. (2002). IT alignment in small firms. *European Journal of Information Systems*, 11, 108–127.
Hutchins, J. (1996). The state of machine translation in Europe. In *Expanding MT horizons: proceedings of the Second Conference of the Association for Machine Translation in the Americas* (pp. 198–205). AMTA.
Hutchins, J. (1998). The origins of the translator's workstation. *Machine Translation*, 13(4), 287–307.
Hutchins, J. (1999). The development and use of machine translation systems and computer-based translation tools. *In International Symposium on Machine Translation and Computer Language Information Processing*.
Hutchins, J. (2000a). Compendium of translation software: commercial machine translation systems and computer-aided translation support tools. *Geneva: European Association for Machine Translation*. (on behalf of International Association for Machine Translation).
Hutchins, J. (2000b). The IAMT Certification initiative and defining translation system categories. In *Fifth EAMT Workshop: Harvesting existing resources*. EAMT.
Hutchins, J. (2001a). Machine translation and human translation: in competition or in complementation? *International Journal of Translation*, 13(1-2), 5–20.
Hutchins, J. (2001b). Machine translation over fifty years. *Histoire Epistémologie Langage*, 23(1), 7–31.
Hutchins, J. (2002). *The state of machine translation in Europe and future prospects*. HLT Central, January 2002. Available at: http://www.hutchinsweb.me.uk/HLT-2002.pdf.

Hutchins, J. (2005a). *Computer-based translation systems and tools.* [Revision 2005]. British Computer Society state of the art reports: natural language translation. Available at: http://www.hutchinsweb.me.uk/BCS-NLT-2005.pdf.

Hutchins, J. (2005b). Current commercial machine translation systems and computer-based translation tools: system types and their uses. *International Journal of Translation, 17* (1–2), 5–38.

Hutchins, J. (2014). *The history of machine translation in a nutshell.* Available at: http://www.mt-archive.info/10/Hutchins-2014.pdf.

Hutchins, J., & Somers, H. (1992). *An introduction to machine translation.* London: Academic Press, Ltd.

IBM (1981). *A management system for the information business.* NY: IBM.

Igbaria, M., Zinatelli, N., & Cavaye, A. L. M. (1998). Analysis of information technology success in small firms in New Zealand. *International Journal of Information Management, 18*(2), 103–119.

Irani, Z., & Love, P. E. D. (2001a). Information systems evaluation: past, present and future. *European Journal of Information Systems, 10*, 183–188.

Irani, Z., & Love, P. E. D. (2001b). The propagation of technology management taxonomies for evaluating investments in information systems. *Journal of Management Information Systems, 17*(3), 161–177.

Irani, Z., & Love, P. E. D. (2002). Developing a frame of reference for ex-ante IT/IS investment evaluation. *European Journal of Information Systems, 11*(1), 74–82.

Irani, Z., Sharif, A. M., & Love, P. E. D. (2001). Transforming failure into success through organisational learning: an analysis of a manufacturing information system. *European Journal of Information Systems, 10*(1), 55–66.

Ismail, N. A. (2004). *AIS alignment in small and medium sized firms.* Ph. D. Thesis. Loughborough: Loughborough University.

ISO/TC 37/SC 3 (2008). ISO 30042:2008 Systems to manage terminology, knowledge and content – TermBase eXchange (TBX). Available at: *http://www.iso.org/iso/iso_catalogue/catalogue_tc/catalogue_detail.htm?csnumber=45797.*

Jaekel, G. (2000). Terminology management at Ericsson. In R. C. Sprung (Ed.), *Translating into success: cutting-edge strategies for going multilingual in a global age.* Amsterdam (pp. 159–171). Philadelphia: John Benjamins Publishing Company.

Jakobson, R. (1959). On linguistic aspects of translation. In R. A. Brower (Ed.), *On translation* (pp. 232–239). Cambridge, MA: Harvard University Press.

Jakobson, R. (1971). *Selected writings: word and language.* The Hague: Walter de Gruyter.

Jarche, H. (2008). Skills 2.0: Web 2. 0 gives learning professionals an aptitude adjustment. *T+D, 62*(4), 22–24.

Jessup, L. M., & Valacich, J. S. (2005). *Information systems today* (2nd ed.). New Jersey: Prentice Hall PTR.

JISC (2002). *The Big Blue: final report,* Available at: http://www.library.mmu.ac.uk/bigblue/finalreport.html [Accessed January 20, 2014].

Johannisson, J., & Sundin, O. (2007). Putting discourse to work: information practices and the professional project of nurses. *The Library Quarterly, 77(2).*

Johnston, B., & Webber, S. (2003). Information literacy in higher education: a review and case study. *Studies in Higher Education, 28*(3), 335.

Joint Information Systems Committee (2003). *JISC Circular 6/03 (Revised). An invitation for expressions of interest to establish a new Digital Curation Centre for research into and support of the curation and preservation of digital data and publications.* Available at: http://www.jisc.ac.uk/uploaded_documents/6-03%20Circular.doc.

Joscelyne, A. (2003). Europe's language industry in 2003. *The LISA Newsletter: Globalization Insider, XII* (1.1).
Kagan, A., Lau, K., & Nusgart, K. R. (1990). Information system usage within small business firms. *Entrepreneurship Theory and Practice, 14*(3), 25–37.
Kaiser, H. F. (1970). A second-generation Little Jiffy. *Psychometrika, 35*, 401–415.
Kaiser, H. F. (1974). Little Jiffy, Mark IV. *Educational and Psychology Measurement, 34*, 111–117.
Kanavos, P., & Kartsaklis, D. (2010). Integrating machine translation with translation memory: a practical approach. In *Proceedings of the Second Joint EM+/CNGL Workshop "Bringing MT to the User: Research on Integrating MT in the Translation Industry" (JEC '10)*. Denver, CO, pp. 11–20. Available at: https://www.cs.ox.ac.uk/files/5267/JEC-2010-Kanavos.pdf [Accessed February 26, 2014].
Kastelle, T. (2013). Hierarchy is overrated. *Harvard Business Review*. Available at: http://blogs.hbr.org/2013/11/hierarchy-is-overrated/ [Accessed April 18, 2014].
Kay, M. (1997). The proper place of men and machines in language translation. *Machine Translation, 12*(1–2), 3–23.
Kellow, J. T. (2005). Exploratory factor analysis in two measurement journals: hegemony by default. *Journal of Modern Applied Statistical Methods, 4*(1), 283–287.
Kelly, D. (2002). Un modelo de competencia traductora: bases para el diseño curricular. *Puentes: Hacia nuevas investigaciones en la mediación intercultural, 1*, 9–20.
Kelly, N. (2009). Freelance translators clash with LinkedIn over crowdsourced translation. Available at: *http://www.commonsenseadvisory.com/Default.aspx?Contenttype=ArticleDetAD&tabID=63&Aid=591&moduleId=391* [Accessed August 7, 2013].
Kempis, R., & Ringbeck, J. (1999). *Do IT smart: seven smart rules for superior information technology performance*. NY: Free Press.
Khandwalla, P. N. (1977). *The design of organisations*. New York: Harcourt Brace Jovanovich, Inc.
Kim, J. (1975). Factor analysis. In H. Norman (Ed.), *Statistical package for the social science*. New York: McGraw Hill.
Kim, J., & Mueller, C. W. (1978). *Factor analysis: statistical method and practical issues*. Beverly Hills, CA: Sage University Press.
King, A. (1993). From sage on the stage to guide on the side. *College Teaching, 41*(1), 30–35.
King, J. L., & Kraemer, K. L. (1984). Evolution and organizational information systems: an assessment of Nolan's stage model. *Communications of the ACM, 27*(5), 466–475.
King, M. (1997). Evaluation design: the EAGLES framework. *In Konvens 97 Proceedings*.
King, M. (1998). Workflow, computer aids and organisational issues. In *1998 workshop of the European Association for Machine Translation: Translation technology: integration in the workflow environment*. Geneva: WHO.
King, M., & McAulay, L. (1989). Information technology and the accountant: a case study. *Behaviour and Information Technology, 8*(2), 109–123.
Kinnear, P. R., & Gray, C. D. (2000). *SPSS for Windows made simple*. Hove: Psychology Press.
Kirk, J. (2004). Information and work: extending the roles of information professionals. In *Challenging ideas—Proceedings of the ALIA 2004 Biennial Conference*. Citeseer, pp. 21–4.
Kirton, J., & Barham, L. (2005). Information literacy in the workplace. *The Australian Library Journal, 54*(4), 365–376.
Kole, M. A. (1983). Going outside for MIS implementation. *Information & Management, 6*, 261–268.

Kuhlthau, C. C. (1991). Inside the search process: Information seeking from the user's perspective. *Journal of the American Society for Information Science*, *42*(5), 361–371.

Kuhlthau, C. C., & Tama, S. L. (2001). Information search process of lawyers: a call for "just for me" information services. *Journal of Documentation*, *57*(1), 25–43.

Kvanli, A. H., Pavur, R. J., & Keeling, K. (2002). *Introduction to business statistics: a Microsoft Excel integrated approach*. Thomson South-Western.

Lagoudaki, E. (2008). The value of machine translation for the professional translator. In *Proceedings of the 8th Conference of the Association for Machine Translation in the Americas*. pp. 262–9. Available at: *http://www.amtaweb.org/papers/3.04_Lagoudaki.pdf* [Accessed August 18, 2013].

Lagoudaki, E. (2006). Translation memories survey 2006 Users' perceptions around tm use. In *Proceedings of the ASLIB International Conference Translating and the Computer*. pp. 1–29. Available at: *http://mt-archive.info/Aslib-2006-Lagoudaki.pdf* [Accessed February 18, 2014].

Lambert, D. M., & Harrington, T. C. (1990). Measuring nonresponse bias in customer service mail surveys. *Journal of Business Logistics*, *11*(2), 5–25.

Lange, C. A., & Bennett, W. S. (2000). Combining machine translation with translation memory at Baan. In R. C. Sprung (Ed.), *Translating into success: cutting-edge strategies for going multilingual in a global age* (pp. 203–218). Amsterdam, Philadelphia: John Benjamins Publishing Company.

Langé, J., Gaussier, E., & Daille, B. (1997). Bricks and skeletons: some ideas for the near future of MAHT. *Machine Translation*, *12*(2), 39–51.

Langewis, C. (2002). What is language technology? *Multilingual Computing, (#51 Supplement)*, 4–10.

Lasnier, F. (2000). *Réussir la formation par compétences*. Montreal: Guérin.

Lees, J. D. (1987). Successful development of small business information systems. *Journal of Systems Management*, *38*(8), 32–39.

Lefebvre, et al. (1996). Intangible assets as determinants of advanced manufacturing technology adoption in SMEs. *IEEE Transactions on Engineering Management*, *43*(3), 307–320.

Lefebvre, L. A., & Lefebvre, E. (1988). Computerization of small firms: a study of the perceptions and expectations of managers. *Journal of Small Business & Entrepreneurship*, *5*(5), 48–58.

Lehman, D. R. (1989). *Market research and analysis*. Singapore: Richard D. Irwin Inc.

Lehman, J. A. (1985). *Organizational size and information systems sophistication*, Working Paper #85-18, MIS Research Center, University of Minnesota.

Lewis-Beck, M. S., et al. (2003). *Sage encyclopedia of social science research methods*. Sage Publications Inc.

Li, L. (2009). *Emerging technologies for academic libraries in the digital age*. Oxford: Chandos Publishing.

Lincoln, A. (2011). FYI: TMI: Toward a holistic social theory of information overload. *First Monday*, 16(3). Available at: *http://ojs-prod-lib.cc.uic.edu/ojs/index.php/fm/article/view/3051* [Accessed February 18, 2014].

Lindner, J. R., Murphy, T. H., & Briers, G. E. (2001). Handling nonresponse in social science research. *Journal of Agricultural Education*, *42*(4), 43–53.

Lloyd, A. (2006). Information literacy landscapes: an emerging picture. *Journal of Documentation*, *62*(5), 570–583.

Lloyd, A. (2010). *Information literacy landscapes: information literacy in education, workplace and everyday contexts*. Oxford: Chandos Publishing.

Lloyd, A. (2003). Information literacy the meta-competency of the knowledge economy? An exploratory paper. *Journal of Librarianship and Information Science, 35*(2), 87–92.

Lloyd, A. (2007). Recasting information literacy as sociocultural practice: implications for library and information science researchers. *Information Research, 12,* 1–13.

Lloyd, A. (2011). Trapped between a rock and a hard place: what counts as information literacy in the workplace and how is it conceptualized? *Library Trends, 60*(2), 277–296.

Lloyd, A. (2004). Working (in) formation: conceptualizing information literacy in the workplace, Working (in) formation: conceptualizing information literacy in the workplace. Available at: *http://www.voced.edu.au/content/ngv27462* [Accessed April 12, 2014].

Locke, N. A. (2005). In-house or Freelance? A Translator's View. *MultiLingual Computing & Technology, 16*(1), 19-21.

Lockwood, R. (2000). Machine translation and controlled authoring at Caterpillar. In R. C. Sprung (Ed.), *Translating into success: cutting-edge strategies for going multilingual in a global age* (pp. 187–202). Amsterdam, Philadelphia: John Benjamins Publishing Company.

Loertscher, D. V., & Woolls, B. (2002). *Information literacy: a review of the research: a guide for practitioners and researchers*. San Jose (Calif.): Hi Willow Research and Pub.

Lommel, A. (2002). *LISA 2002 translation memory survey: translation memory and translation memory standards*. LISA.

Lommel, A. (2004). *LISA 2004 translation memory survey: translation memory and translation memory standards*. LISA.

Magal, S. R., & Lewis, C. D. (1995). Determinants of information technology success in small businesses. *Journal of Computer Information Systems, 35*(3), 75.

Mahmood, M. A., & Becker, J. D. (1985). Impact of organizational maturity on user satisfaction with information systems. In *21st Annual Computer Personnel Research Conference* (pp. 134–151). ACM.

Malone, S. C. (1985). Computerizing small business information systems. *Journal of Small Business Management, 23*(2), 10–16.

Mann, C., & Stewart, F. (2000). *Internet communication and qualitative research*. London: Sage.

Markless, S., & Streatfield, D. R. (2007). Three decades of information literacy: redefining the parameters. In *Change and challenge: information literacy for the 21st century* (pp. 15–36). Adelaide: Auslib Press.

Marland, M. (1981). *Information skills in the secondary curriculum the recommendations of a working group sponsored by the British Library and the Schools Council*. Schools Council.

Martin, C. J. (1989). Information management in the smaller business: the role of the top manager. *International Journal of Information Management, 9*(3), 187–197.

Massey, G., & Ehrensberger-Dow, M. (2011). Investigating information literacy: a growing priority in translation studies. *Across Languages and Cultures, 12*(2), 193–211.

Mata Pastor, M. (2008). Formatos libres en traducción y localización. In O. Díaz Fouces, & M. García González (Eds.), *Traducir (con) software libre* (pp. 75–122). Granada: Comares.

McFarlan, F. W., McKenney, J. L., & Pyburn, P. (1983). The information archipelago – plotting a course. *Harvard Business Review, 61*(1), 145–156.

McKay, J., & Marshall, P. (2001). The IT evaluation and benefits management life cycle. In W. V. Grembergen (Ed.), *Information technology evaluation methods and management* (pp. 44–56). London: Idea Group Publishing.

Melby, A. (1998). Eight types of translation technology. In *ATA 39th Annual Conference*. ATA.

Melby, A. (1982). Multi-level translation aids in a distributed system. In J. Horecký (Ed.), *Proceedings of COLING 82* (pp. 215–220). North Holland Publishing Company.

Melby, A. (1992). The translator workstation. In J. Newton (Ed.), *Computers in translation: a practical appraisal* (pp. 147–165). London: Routledge.
Melby, A.K. (2007). MT+ TM+ QA: the future is ours. *Tradumàtica: traducció i tecnologies de la informació i la comunicació, (4)*. Available at: http://www.raco.cat/index.php/Tradumatica/article/viewArticle/56004/0 [Accessed August 18, 2013].
Melby, A.K., et al. (2012). Reliably assessing the quality of post-edited translation based on formalized structured translation specifications. In *AMTA 2012 Workshop on post-editing technology and practice (WPTP 2012)*. San Diego, USA, pp. 31–40. Available at: http://mt-archive.info/AMTA-2012-Melby.pdf [Accessed August 18, 2013].
Melby, A. K. (2012). Terminology in the age of multilingual corpora. *Journal of Specialized Translation, 18*, 7–29.
Mell, P., & Grance, T. (2009). The NIST definition of cloud computing. *National Institute of Standards and Technology, 53*(6), 50.
Melville, N., Kraemer, K., & Gurbaxani, V. (2004). Review: information technology and organizational performance: an integrative model of IT business value. *MIS quarterly, 28*(2), 283–322.
Mesipuu, M. (2012). Translation crowdsourcing and user-translator motivation at Facebook and Skype. *Translation Spaces, 1*(1), 33–53.
Miles, M. B., & Huberman, A. M. (1994). *Qualitative data analysis: an expanded sourcebook* (2nd ed.). California: Sage.
Miller, D., & Droge, C. (1986). Psychological and traditional determinants of structure. *Administrative Science Quarterly, 31*(4), 539–560.
Molnar, A. (2014). *Information overload*. Available at: http://www.aiim.org/community/blogs/expert/Information-Overload [Accessed February 18, 2014].
Montalt i Resurrecció, V. (2009). La consulta documental y humana aplicada a la traducción médica: Reflexiones en torno a la práctica profesional y a la pedagogía. In M. Pinto et al.,(Eds.). *Documentación aplicada y Espacio Europeo de Educación Superior* (pp. 171–186). Madrid: Arco Libros.
Montazemi, A. R. (1988). Factors affecting information satisfaction in the context of the small business environment. *MIS Quarterly, 12*(2), 239–256.
Moore, G. C., & Benbasat, I. (1991). Development of an instrument to measure the perceptions of adopting an information technology innovation. *Information Systems Research, 2*(3), 192–222.
Mumford, E., et al. (1985). *Research methods in information systems*. Amsterdam: North-Holland Publishing Co.
Mutch, A. (1997). Information literacy: an exploration. *International Journal of Information Management, 17*(5), 377–386.
Nida, E. A., & Taber, C. R. (1969). *The theory and practice of translation*. Leiden: EJ Brill.
Nielsen, H. J., & Hjørland, B. (2014). Curating research data: the potential roles of libraries and information professionals. *Journal of Documentation, 70*(2), 221–240.
Nolan, R. L. (1973). Managing the computer resource: a stage hypothesis. *Communications of the ACM, 16*(7), 399–405.
Nolan, R. L. (1979). Managing the crisis in data processing. *Harvard Business Review, 57*(2), 115–126.
Nord, B. (2002). *Hilfsmittel beim Übersetzen*. Lang.
O'Hagan, M., & Ashworth, D. (2002). *Translation-mediated communication in a digital world facing the challenges of globalization and localization*. Clevedon: Multilingual Matters.
O'Sullivan, C. (2002). Is information literacy relevant in the real world? *Reference Services Review, 30*(1), 7–14.

Oman, J. N. (2001). Information literacy in the workplace. *Information Outlook*, 5(6), 32.
ONS (2003). *Labour Force Survey*. Available at: http://www.statistics.gov.uk/STATBASE/ Source.asp?LinkBtn.x=42&LinkBtn.y=10&LinkBtn=Show+Links&ComboState=Show+Links&More=Y&vlnk=358 [Accessed June 2, 2004].
Oppenheim, A. N. (1992). *Questionnaire design, interviewing and attitude measurement*. London: Printer Publishers.
OPTIMALE (2012). *Draft Synthesis Report Wp 5.2 (Domain Specialisation)*, Salamanca: OPTIMALE (Optimising Professional Translator Training in a Multilingual Europe). Available at: http://www.translator-training.eu/attachments/article/37/Synthesis%20Report_Eng.pdf.
Orlikowski, W. J., & Baroudi, J. J. (1991). Studying information technology in organizations: research approaches and assumptions. *Information Systems Research*, 2(1), 1–28.
PACTE group (2005). Investigating translation competence: conceptual and methodological issues. *Meta: Journal des traducteurs*, 50(2), 609.
PACTE group (2009). Results of the validation of the PACTE translation competence model: acceptability and decision making. *Across Languages and Cultures*, 10(2), 207–230.
Palomares Perraut, R. & Pinto Molina, M. (2000). Aproximación a las necesidades, hábitos y usos documentales del traductor. *Terminologie et traduction*(3), 98–124.
Palvia, P. C. (1996). A model and instrument for measuring small business user satisfaction with information technology. *Information & Management*, 31(3), 151–163.
Palvia, P. C., Means, D. B., & Jackson, W. M. (1994). Determinants in computing in very small businesses. *Information & Management*, 27(3), 161–174.
Palvia, P. C., & Palvia, S. C. (1999). An examination of the IT satisfaction of small-business users. *Information & Management*, 35, 127–137.
Pavlovich, H. (1999). Director's notes. *The Linguist*, 38, 37.
Peppard, J., & Ward, J. (2004). Beyond strategic information systems: towards an IS capability. *The Journal of Strategic Information Systems*, 13(2), 167–194.
Pettigrew, A. M., et al. (2003). *Innovative forms of organizing: international perspectives*. SAGE.
Piccoli, G., & Ives, B. (2005). Review: IT-dependent strategic initiatives and sustained competitive advantage: a review and synthesis of the literature. *MIS Quarterly*, 29(4), 747–776.
Pinto, M., et al. (2014). Assessing information competences of translation and interpreting trainees: A study of proficiency at Spanish universities using the InfoliTrans Test. *Aslib Journal of Information Management*, 66(1), 77–95.
Pinto, M. (1999). Competencias del traductor de textos literarios desde la perspectiva documental. *Terminologie et traduction*(3), 99–111.
Pinto, M., & Sales, D. (2007). A research case study for user-centred information literacy instruction: information behaviour of translation trainees. *Journal of Information Science*, 33(5), 531–550.
Pinto, M., & Sales, D. (2008). INFOLITRANS: a model for the development of information competence for translators. *Journal of Documentation*, 64(3), 413–437.
Pinto, M., & Sales, D. (2008). Towards user-centred information literacy instruction in translation: the view of trainers. *The Interpreter and Translator Trainer*, 2(1), 47–74.
Popova, M. (2011). *In a new world of informational abundance, content curation is a new kind of authorship*. Nieman Journalism Lab. Available at: http://www.niemanlab.org/2011/06/maria-popova-in-a-new-world-of-informational-abundance-content-curation-is-a-new-kind-of-authorship/ [Accessed August 22, 2013].
Porter, M. E. (1979). How competitive forces shape strategy, *Harvard Business Review* 57, 137–146.

Porter, M. E. (2008). The five competitive forces that shape strategy. *Harvard Business Review*, *86*(1), 25–40.
Porter, M. E., & Millar, V. E. (1985). How information gives you competitive advantage. *Harvard Business Review, July–August, 1985*, 149–160.
Premkumar, G., & King, W. R. (1994). Organizational characteristics and information systems planning: an empirical study. *Information Systems Research*, *5*(2), 75–109.
Proudlock, M., Phelps, B., & Gamble, P. (1999). IT adoption strategies: best practice guidelines for professional SMEs. *Journal of Small Business and Enterprise Development*, *6*(3), 240–252.
Pym, A. (2004). *The moving text: localization, translation, and distribution*. Amsterdam, Philadelphia: John Benjamins.
Quah, C. K. (2006). *Translation and technology*. New York: Palgrave Macmillan.
Ragin, C. C. (1987). *The comparative method: moving beyond qualitative and quantitative strategies*. Berkeley: University of California Press.
Raymond, L. (1987). An empirical study of management information systems sophistication in small business. *Journal of Small Business & Entrepreneurship*, *5*(1), 38–47.
Raymond, L. (1988). La sophistication des systèmes d'information en contexte PME: une approche par le portefeuille d'applications. *MIS Quarterly*, *5*(2), 32–39.
Raymond, L. (1989). Management information systems: problems and opportunities. *International Small Business Journal*, *7*(4), 44–53.
Raymond, L. (1985). Organizational characteristics and MIS success in the context of small business. *MIS Quarterly*, *9*(1), 37–52.
Raymond, L. (1990). Organizational context and information systems success: a contingency approach. *Journal of Management Information Systems*, *6*(4), 5–18.
Raymond, L., & Paré, G. (1992). Measurement of information technology sophistication in small manufacturing businesses. *Information Resources Management Journal, Spring*, 4–16.
Raymond, L., Paré, G., & Bergeron, F. (1995). Matching information technology and organizational structure: an empirical study with implications for performance. *European Journal of Information Systems*, *4*, 3–16.
Reuther, U. (1999). *LETRAC Final Report. Deliverable D5*.
Rico Pérez, C. (2002). Translation and project management. Translation Journal, 6(4.).
Rinsche, A. (2000). Computer-assisted business process management for translation and localisation companies. In *Fifth EAMT Workshop: Harvesting existing resources*.
Rockart, J. F. (1983). The information era. In *IBM seminar on management decisions in information systems*. Finland: IBM.
Rogers, E. M. (1995). *Diffusion of innovations*. New York: The Free Press.
Rosenberg, V. (2002). *Information literacy and small business. White paper prepared for UNESCO, the US National Commission on Libraries and Information Science, and the National Forum on Information Literacy*. Available at: www.nclis.gov/libinter/infolitconf&meet/papers/rosenberg-fullpaper.pdf.
Rotman, D., et al. (2012). Supporting content curation communities: the case of the Encyclopedia of Life. *Journal of the American Society for Information Science and Technology*, *63*(6), 1092–1107.
Roxburgh, A. (2004). Translating is EU's new boom industry,.
Rubio, E., et al. (2011). eProfessional: from PLE to PLWE. In *The PLE Conference 2011*. Southampton, UK, pp. 1–10. Available at: http://journal.webscience.org/597/ [Accessed April 23, 2014].
Saarinen, T. (1996). An expanded instrument for evaluating information systems success. *Information & Management*, *31*(2), 103–118.

Saarinen, T. (1989). Evolution of information systems in organizations. *Behaviour and Information Technology*, *8*(5), 387–398.

Sainz-Aloy, A., & Soy-Aumatell, C. (2011). Gestión eficiente del correo electrónico: una experiencia corporativa. *El Profesional de la Informacion*, *20*(5), 571–576.

Sales, D. (2008). Towards a student-centred approach to information literacy learning: A focus group study on the information behaviour of translation and interpreting students. *Journal of Information Literacy*, *2*(1), 41–60.

Sales, D., & Pinto, M. (2011). The professional translator and information literacy: perceptions and needs. *Journal of Librarianship and Information Science*, *43*(4), 246–260.

Sales, D., Pinto, M., & Granell, X., Forthcoming. The documentary process in translation: a case study of information behaviour among translation trainees.

Saunders, C. S., & Keller, R. T. (1983). A study of maturity of the information system function task characteristics and interdepartmental communication. In *Proceedings of the 4th annual ICIS* (pp. 111–124). Houston.

Schäffner, C. (2000). *Translation in the global village*. Clevedon: Multilingual Matters.

Scheuermann, L., & Taylor, G. (1997). Netiquette. *Internet Research*, *7*(4), 269–273.

Schmitt, P. A. (1998). Berufsbild. In M. Snell-Hornby et al.,(Eds.), *Handbuch translation* (pp. 1–5). Stauffenburg: Tübingen.

Schumpeter, J. A. (1934). The theory of economic development. *Harvard Economic Studies*, 46.

SCONUL (2011). *The SCONUL seven pillars of information literacy: core model for higher education*. Available at: http://www.sconul.ac.uk/sites/default/files/documents/coremodel.pdf.

Serafeimidis, V., & Smithson, S. (1996). The management of change for information systems evaluation practice: experience from a case study. *International Journal of Information Management*, *16*(3), 205–217.

Shadbolt, D. (2003). Translation processes and tools. *Multilingual Computing & Technology, #53 Supplement Guide to Translation*, *2003*, 4–8.

Shannon, C. E., & Weaver, W. (1949). *The mathematical theory of communication*. University of Illinois Press.

Shea, V., & Shea, C. (1994). *Netiquette*. Albion Books.

Shields, M. (1999). Slaves to the computer. *Institute of Translation and Interpreting Bulletin, October, 1999*, 4–5.

Siemens, G. (2005). Connectivism: a learning theory for the digital age. *International Journal of Instructional Technology and Distance Learning*, *2*(1), 3–10.

Slocum, J. (1988). *Machine translation systems*. Cambridge: Cambridge University Press.

Smith, D., & Tyldesley, D. (1986). Translation practices report. *External report, Digital Equipment Corporation*.

Soh, C. P. P., Yap, C. S., & Raman, K. S. (1992). Impact of consultants on computerization success in small businesses. *Information & Management*, *22*(4), 309–319.

Somers, H. (2003a). *Computers and translation: a translator's guide*. Amsterdam, Philadelphia: John Benjamins.

Somers, H. (2003b). The translator's workstation. In H. Somers (Ed.), *Computers and translation: a translator's guide* (pp. 13–30). Amsterdam, Philadelphia: John Benjamins.

Somers, H. (2003c). Translation memory systems. In H. Somers (Ed.), *Computers and translation: a translator's guide* (pp. 31–47). Amsterdam, Philadelphia: John Benjamins.

Sproull, N. L. (1988). *Handbook of research methods: a guide for practitioners and students in the social sciences*. New Jersey: The Scarecrow Press.

Sprung, R. C. (2000). *Translating into success: cutting-edge strategies for going multilingual in a global age*. Amsterdam, Philadelphia: John Benjamins.

Stair, R., & Reynolds, G. (2011). *Principles of information systems* (10th ed.). Cengage Learning.

Stair, R. M., Crittenden, W. F., & Crittenden, V. L. (1989). The use, operation and control of the small business computer. *Information & Management, 16*(3), 125–130.
Stewart, D. W. (1981). The application and misapplication of factor analysis in marketing research. *Journal of Marketing Research, 18*, 51–62.
Tapscott, D. (2004). The engine that drives success: the best companies have the best business models because they have the best IT strategies. *CIO Magazine*.
Taravella, A., & Villeneuve, A. O. (2013). Acknowledging the needs of computer-assisted translation tools users. *the human perspective in human-machine translation*, 62–74.
TAUS (2009). *LSPs in the MT loop: current practices, future requirements*. Available at: https://www.taus.net/reports/lsps-in-the-mt-loop-current-practices-future-requirements [Accessed August 5, 2013].
Thong, J. Y. L. (1999). An integrated model of information systems adoption in small businesses. *Journal of Management Information Systems, 15*(4), 187–214.
Thong, J. Y. L., Yap, C. -S., & Raman, K. S. (1996). Top management support, external expertise and information systems implementation in small businesses. *Information Systems Research*, 7(2), 248–267.
Toffler, A. (1970). *Future shock*. New York: Random House.
Tunick, L. (2003). Finding a cost-effective translation solution. *Multilingual Computing & Technology, #53 Supplement Guide to Translation*, 13–15.
Tuominen, K., Savolainen, R., & Talja, S. (2005). Information literacy as a sociotechnical practice. *The Library Quarterly, 75*(3), 329–345.
Turner, J. S. (1982). Firm size, performance, and computer use. In *3rd International Conference on Information Systems*. Ann Arbor, Michigan, pp. 109–20.
UNESCO (2006). *High-level colloquium on information literacy and lifelong learning Bibliotheca Alexandrina, Alexandria, Egypt - Report of Meeting*. Available at: http://archive.ifla.org/III/wsis/BeaconInfSoc-es.html [Accessed January 9, 2014].
Varona, L. (2002). El traductor ante la micro y pequeña empresa PYME. In A. Alcina Caudet, & S. Gamero Pérez (Eds.), *La traducción científico-técnica y la terminología en la sociedad de la información* (pp. 201–206). Castellón: Publicacions de la Universitat Jaume I.
Virkus, S. (2012). Information literacy from the policy and strategy perspective. *Nordic Journal of Information Literacy in Higher Education, 4*(1), 16–36.
Virkus, S. (2003). Information literacy in Europe: a literature review. *Information research, 8*(4).
Virkus, S. (2013). Information literacy in Europe: ten years later. In *Worldwide commonalities and challenges in information literacy research and practice*. Springer, pp. 250–7. Available at: http://link.springer.com/chapter/10.1007/978-3-319-03919-0_32 [Accessed January 22, 2014].
Voorsluys, W., Broberg, J., & Buyya, R. (2010). Introduction to cloud computing. In R. Buyya, J. Broberg, & A. M. Goscinski (Eds.), *Cloud computing: principles and paradigms* (pp. 1–44). New York, NY, USA: John Wiley & Sons.
Vygotsky, L. S. (1978). *Mind and society: the development of higher mental processes*. Cambridge, MA: Harvard University Press.
Wade, M., & Hulland, J. (2004). Review: The resource-based view and information systems research: Review, extension, and suggestions for future research. *MIS quarterly, 28*(1), 107–142.
Webb, L. E. (2000). *Advantages and disadvantages of translation memory: a cost/benefit analysis* M.A. thesis. Monterey, California: Monterey Institute of International Studies.
Weill, P., & Baroudi, J. (1990). *An empirical investigation of the relationship between firm performance and system success*. University of Melbourne. Graduate School of Management.
Weill, P., & Olson, M. H. (1989). Managing investment in information technology: mini case examples and implications. *MIS Quarterly, 13*(1), 3–17.

Weßel, R. (1995). Trados: MultiTerm for Windows user report In TAMA'94: *Terminology in Advanced Microcomputer Applications*. In *3rd TermNet Symposium* (pp. 87–105). TermNet.
Whittaker, S. (2011). Personal information management: from information consumption to curation. *Annual Review of Information Science and Technology*, 45(1), 1–62.
Whyman, E. K., & Somers, H. L. (1999). Evaluation metrics for a translation memory system. *Software-Practice and Experience*, 29(14), 1265–1284.
Winkin, Y. (1984). La nueva comunicación. Editorial Kairós.
Wright, S. E., & Budin, G. (Eds.). (2001). *Handbook of terminology management: Volume 2: Application-oriented terminology management*. Amsterdam, Philadelphia: John Benjamins.
Yang, J., & Lange, E.D. (1998). SYSTRAN on AltaVista a user study on real-time machine translation on the Internet. In *Machine translation and the information soup*. Springer, pp. 275–85. Available at: *http://link.springer.com/chapter/10.1007/3-540-49478-2_25* [Accessed February 26, 2014].
Yap, C. S., Soh, C. P. P., & Raman, K. S. (1992). Information systems success factors in small business. *OMEGA - International Journal of Management Science*, 20(5/6), 597–609.
Ye, N. (2003). *The handbook of data mining*. Lawrence Erlbaum Associates.
Yin, R. K., & Campbell, D. T. (2002). *Case study research: design and methods*. Newbury Park, Calif: Sage Publications Inc.
Yoganarasimhan, H. (2013). *The value of reputation in an online freelance marketplace*, Rochester, NY: Social Science Research Network.
Yuste Rodrigo, E. (2013). La posedición en el flujo de producción de contenido multilingüe: tendencias, actantes e implicaciones tecnológicas. *Revista Tradumàtica: tecnologies de la traducció*, 0(10), 157–165.
Zerfass, A. (2002). Evaluating translation memory systems. In *Language resources for translation work and research*. pp. 49–53. Available at: http://lrec.elra.info/proceedings/lrec2002/pdf/ws8.pdf#page=56 [Accessed February 20, 2014].
Zetzsche, J. (2006). Translation tools come full circle. *MultiLingual*, 17(1), 41.
Zetzsche, J., & Melby, A.K. (2013). *Jost Zetzsche interviews Prof. Alan K. Melby | The Big Wave*. Available at: http://thebigwave.it/technology-here-and-now/melby-interview/ [Accessed February 25, 2014].
Zmud, R. W., Boynton, A. C., & Jacobs, G. C. (1987). An examination of managerial strategies for increasing information technology penetration in organizations. In *Proceedings of the 8th International Conference on Information Systems* (pp. 24–44). Pittsburg.
Zurkowski, P. G. (1974). *The information service environment relationships and priorities*. Related paper No. 5. Available at: *http://files.eric.ed.gov/fulltext/ED100391.pdf* [Accessed February 12, 2014].

Appendix 1

The Business School
Loughborough University
Loughborough, LE11 3TU
Telephone: 01509 228842
Fax: 01509 223960
E-mail: j.granell-zafra@lboro.ac.uk

Translators in the 21st century: a study of skills, software and strategies

Approximate time for completion: 10-15 mins.

If you wish to make comments on any question, please use the space provided on the back cover

---

ALL RESPONSES WILL BE TREATED IN THE STRICTEST CONFIDENCE

Your answers are very important to the accuracy of this study. Please return this questionnaire at your earliest convenience using the self-addressed envelope provided.

Thank you for your cooperation.

Please circle an appropriate number, or tick ☑ the relevant boxes, or write your answer as appropriate

SECTION A: TRANSLATOR PROFILE

Please provide some background information. *(Please tick the appropriate box or fill in the required data)*

1. **Please indicate your age range**
   - ☐ 20-29 ☐ 30-39 ☐ 40-49 ☐ 50-59 ☐ 60+

2. **Please indicate your gender**
   - ☐ Male ☐ Female

3. **Please indicate your highest educational level**
   - ☐ University – bachelor ☐ University – masters ☐ University – doctorate
   - ☐ Other (please specify): ....

4. **Please tell us which of the following translation qualifications you hold** *(tick all that apply)*
   - ☐ University (bachelor) in translation / translation studies
   - ☐ Postgraduate degree in translation / translation studies
   - ☐ Translation diploma (e.g. DipTrans IoL)
   - ☐ Other (please specify): ....

5. **Please indicate your role** *(tick all that apply)*
   - ☐ Freelance translator
   - ☐ In-house translator
   - ☐ Manager of a translation company
   - ☐ Other (please specify): ....

   PLEASE NOTE:
   If you are **not actively involved** in translation work at present, please tick here ☐, do not proceed with the rest of the questionnaire, but **return** it in the self-addressed envelope. **Thank you for your time.**

6. **In which year did you start working as a translator**
   [_____]

7. **Please indicate your membership status (if any) in the following professional institutes** *(tick all that apply)*

IoL (Institute of Linguists):	☐ FIL	☐ MIL	☐ AIL	☐ Student
ITI (Institute of Translation & Interpreting):	☐ Member	☐ Associate	☐ Student Associate	
ATC (Association of Translation Companies):	☐ Full member	☐ Associate	☐ Overseas membership	

   Other (please specify): ....

8. **Please indicate the approximate average number of words that you translate each week**
   [_____] words per week

9. **Please indicate the approximate average number of hours that you dedicate to translation-related tasks each week**
   [_____] hours per week

10. Please indicate how often you employ the services of:

	Never	For a few translation assignments	For some translation assignments	For most translation assignments	For all translation assignments
Proof-readers	☐	☐	☐	☐	☐
Revisers	☐	☐	☐	☐	☐
Other translators	☐	☐	☐	☐	☐
Clerical/Administrative personnel	☐	☐	☐	☐	☐

11. Please indicate any additional services that you provide *(tick all that apply)*

☐ Translation project management ☐ Language training courses/tutorials
☐ Software localisation ☐ Linguistic consultancy
☐ Website localisation ☐ Subtitling/Dubbing
☐ Other (please specify): ....

12. Please tell us which language pairs you translate *(your 'top three' only)*

FROM .... TO ....
FROM .... TO ....
FROM .... TO ....

13. Please indicate the subject areas you translate *(tick all that apply)*

☐ Financial translation ☐ Technical translation ☐ Business/Commerce translation
☐ Legal translation ☐ Scientific translation ☐ Literary translation
Other (please specify): ....

SECTION B: INFORMATION TECHNOLOGY (IT) USAGE

14. Please indicate how you acquired your IT skills *(tick all that apply)*

☐ Self taught
☐ Professional training courses
☐ Attending workshops and seminars run by professional institutes
☐ University/College course
☐ Attending IT modules on University degree programme
☐ Other (please specify): ....

15. Do you have any formal IT qualifications?

☐ Yes ☐ No

If YES, please indicate the type of qualification

☐ University degree in computing/IT
☐ Individual IT modules on University degree programme
☐ School/College qualification (e.g. GCSE, A-Level)
☐ Professional certificate (e.g. European Computer Driving Licence)
☐ Other (please specify): ....

16. Which type of network do you use?

☐ Internet dial-up connection ☐ Local Area Network (LAN)
☐ Internet broadband connection (please, specify speed: KB)
☐ Other (please specify): ....

17. Please provide an indication of your familiarity and experience with each of the following SOFTWARE APPLICATIONS and then indicate which ones you are currently using
(please tick all the boxes that apply and circle only one number per line)

Type of software application	FAMILIARITY AND WORKING KNOWLEDGE				USAGE
	Not familiar	Familiar, but with no working knowledge	Familiar, with some working knowledge	Familiar, with extensive working knowledge	Currently using
Word processing package (e.g. Microsoft Word, Wordperfect)	1	2	3	4	☐
Spreadsheet package (e.g. Microsoft Excel, Lotus 1-2-3)	1	2	3	4	☐
Database package (e.g. Microsoft Access, FileMaker, FoxPro)	1	2	3	4	☐
Computer-based accounting application (e.g. Sage, Ms Money, Lotus Organizer)	1	2	3	4	☐
Desktop Publishing application (e.g. QuarkXpress, PageMaker, Publisher)	1	2	3	4	☐
Web publishing application (e.g. Dreamweaver, FrontPage, GoLive)	1	2	3	4	☐
Graphics applications (e.g. Photoshop, Paint Shop Pro, Fireworks)	1	2	3	4	☐
Information retrieval and Optical Character Recognition (OCR) tools (e.g. Search & Replace, ht://Dig, Omnipage)	1	2	3	4	☐
Groupware applications (e.g. Lotus Notes, Novell Groupwise)	1	2	3	4	☐
Project and Workflow Management software (e.g. Ms Project, STAR Proactive GMS)	1	2	3	4	☐
Terminology management applications (e.g. MultiTerm, Lingo, Déjà Vu TermWatch)	1	2	3	4	☐
Machine Translation (MT) applications (e.g. Reverso Pro, Systran, Telegraph)	1	2	3	4	☐
Computer-Assisted Translation (CAT) (e.g. Trados Workbench, Déjà Vu, SDLX)	1	2	3	4	☐
Localisation applications (e.g. Alchemy Catalyst, Passolo)	1	2	3	4	☐
Other:	1	2	3	4	☐

Appendix 1

SECTION C: YOUR INTERNET USAGE AND YOUR TRANSLATION ACTIVITIES

18. Do you have your own web site to promote your translation services?

☐ Yes ☐ No

19. Please provide an indication of your familiarity and experience with each of the following INTERNET-BASED RESOURCES and then indicate which ones you are currently using in your translation activities *(please tick all the boxes that apply and circle only one number per line)*

Internet-based resources	FAMILIARITY AND WORKING KNOWLEDGE				USAGE
	Not familiar	Familiar, but with no working knowledge	Familiar, with some working knowledge	Familiar, with extensive working knowledge	Currently using
Online dictionaries & glossaries	1	2	3	4	☐
Multilingual terminology databases	1	2	3	4	☐
Discussion mailing lists	1	2	3	4	☐
Online discussion groups	1	2	3	4	☐
Online translation marketplaces (e.g. Proz)	1	2	3	4	☐
Online machine translation (MT) systems	1	2	3	4	☐
Online encyclopaedia	1	2	3	4	☐
Newspapers & magazines archives	1	2	3	4	☐
Academic journals	1	2	3	4	☐
Electronic databases	1	2	3	4	☐
Online search engines	1	2	3	4	☐
Electronic libraries	1	2	3	4	☐
E-mail	1	2	3	4	☐
FTP (File Transfer Protocol)	1	2	3	4	☐
IRC (Internet Relay Chat)	1	2	3	4	☐
Usenet newsgroups	1	2	3	4	☐
Specialist gateways	1	2	3	4	☐
Other:	1	2	3	4	☐

SECTION D: YOUR IT STRATEGY

Please provide some information about your business strategy and your perceptions towards the use of information technology (IT).

20. Do you have a written business plan?
(A document which contains an analysis of your business' current position, where you would like it to be in the future, and how you plan to get it there)

☐ Yes ☐ No

21. **Please consider the tasks listed below, and indicate first, how IMPORTANT you believe IT to be to each one, and second, the USE you currently make of IT to support each one**
(please circle the numbers that apply)

Importance				IT support for...	Use			
Not important	Less important	Important	Very important		None	Little	Moderate	Extensive
1	2	3	4	Administrative tasks	1	2	3	4
1	2	3	4	Project and document management tasks	1	2	3	4
1	2	3	4	Information retrieval (documentation) tasks	1	2	3	4
1	2	3	4	Translation tasks	1	2	3	4
1	2	3	4	Communication tasks	1	2	3	4

22. **The following statements help us understand your opinions about INFORMATION TECHNOLOGY (IT). Please indicate,** *by circling the most appropriate number on the scale,* **the extent to which you agree with each of the following statements**

	Strongly Disagree	Disagree	Don't Know	Agree	Strongly Agree
Previous experience with computers is necessary for adopting new applications	1	2	3	4	5
Computerisation helps provide higher quality services	1	2	3	4	5
Computerisation brings time saving benefits	1	2	3	4	5
Computerisation would bring more benefits for me if there were a greater level of integration between the various software applications I use	1	2	3	4	5
Computerisation significantly improves my effectiveness as a translator	1	2	3	4	5
Computerisation helps to increase revenue	1	2	3	4	5
Computerisation significantly improves my communication with customers	1	2	3	4	5
Computer applications have failed to meet some of my requirements	1	2	3	4	5
Computerisation creates many problems	1	2	3	4	5
So far, my use of computer applications has been a failure	1	2	3	4	5
I have gained fewer benefits than expected from computerisation	1	2	3	4	5

Appendix 1

23. The following statements help us understand your opinions about COMPUTER-ASSISTED TRANSLATION (CAT) TOOLS (e.g. Trados Workbench, Atril Déjà Vu, SDLX). Please indicate, *by circling the most appropriate number on the scale,* the extent to which you agree with each of the following statements

	Strongly Disagree	Disagree	Don't Know	Agree	Strongly Agree
Previous experience with CAT tools is necessary for adopting a new CAT tool	1	2	3	4	5
CAT tools help provide higher quality services	1	2	3	4	5
CAT tools bring time saving benefits	1	2	3	4	5
CAT tools are well worth their cost	1	2	3	4	5
CAT tools help to increase revenue	1	2	3	4	5
CAT tools significantly improve my effectiveness as a translator	1	2	3	4	5
CAT tools would improve my effectiveness as a translator if they were integrated with other software applications	1	2	3	4	5
CAT tools have failed to meet some of my requirements	1	2	3	4	5
CAT tools create many problems	1	2	3	4	5
So far, my use of CAT tools has been a failure	1	2	3	4	5
I have gained fewer benefits than expected from CAT tools	1	2	3	4	5

24. Below are some pairs of statements about IT STRATEGY. Please indicate for each pair, using the scale below, which statement most closely matches your current position *(please circle only one number per line)*

A						B
I treat each decision about a new IT investment independently	1	2	3	4	5	My decisions about IT investments are guided by a formal IT strategy
I am concerned with using IT to solve short-term problems	1	2	3	4	5	I am concerned with using IT to solve medium to longer-term problems
I am concerned with matching technology to my business needs	1	2	3	4	5	I am concerned with getting the most up-to-date technology
I am concerned with how to better manage my IT resources	1	2	3	4	5	Managing IT is not as critical as managing other non-translation related resources
I am concerned with achieving a greater level of integration of my computer systems	1	2	3	4	5	I am concerned that the majority of my computer systems remain as standalone applications
The primary benefits I seek from IT are improved productivity and efficiency	1	2	3	4	5	Computer systems bring a wide range of benefits including competitive advantage

Please use this space for any comments you wish to make related to this study

....
....
....
....
....
....
....
....

Would you like a copy of the summary of the findings?

☐ Yes   ☐ No

Would you like to participate in other stages of our research?
(e.g. questionnaire, interview)

☐ Yes   ☐ No

If you answered **YES** to either of the above questions, please supply your name, address and e-mail below (or attach a business card).

Alternatively, if you would prefer your responses to remain completely anonymous, you can email Joaquin Granell-Zafra [j.granell-zafra@lboro.ac.uk] stating 'Copy of questionnaire findings' as subject, to request a copy of the findings.

Name	
Address	
E-mail	

ALL RESPONSES WILL BE TREATED IN THE STRICTEST CONFIDENCE

Your answers are very important to the accuracy of this study. Please return this questionnaire at your earliest convenience using the self-addressed envelope provided.

Thank you for your cooperation

Appendix 2. Online survey for CAT tools adopters

Translation Tools in the 21st century

Part A: Terminology management tools

This section contains questions about terminology management tools, i.e. software packages used for creating and managing your own terminology collections. Examples include MultiTerm, Lingo, TermWatch, and StarTerm.

Question 1: Using terminology management tools

Which terminology management tool(s) do you use?

Using the scale provided, please indicate the extent to which you agree or disagree with each of the following statements about terminology management tools.

Terminology management tools...	Strongly Disagree		Neutral		Strongly Agree
1. Enable me to accomplish tasks more quickly.	O	O	O	O	O
2. Improve the quality of work I do.	O	O	O	O	O
3. Make it easier for me to do my job.	O	O	O	O	O
4. Improve my job performance.	O	O	O	O	O
5. Are overall advantageous in my job.	O	O	O	O	O
6. Enhance my effectiveness in my work.	O	O	O	O	O
7. Give me greater control over my work.	O	O	O	O	O
8. Increase my productivity.	O	O	O	O	O
9. Are compatible with the type of translation assignments I undertake.	O	O	O	O	O
10. Fit well with the way I like to work.	O	O	O	O	O
11. Are cumbersome to use.	O	O	O	O	O
12. Require a lot of mental effort.	O	O	O	O	O
13. Are often frustrating.	O	O	O	O	O
14. Do what I want to do easily.	O	O	O	O	O
15. Are easy for me to use.	O	O	O	O	O
16. Were easy for me to learn.	O	O	O	O	O

Question 2: Terminology management tools and the translation sector

Using the scale provided, please indicate the extent to which you agree or disagree with each of the following statements.

Terminology management tools...	Strongly Disagree		Neutral		Strongly Agree
1. My clients expect me to use them.	○	○	○	○	○
2. My use of them is voluntary.	○	○	○	○	○
3. Using them improves my image within the translation sector.	○	○	○	○	○
4. Clients prefer to work with translators who use them.	○	○	○	○	○
5. Translators who use them have a high profile in the translation sector.	○	○	○	○	○
6. Having them is a status symbol among translators.	○	○	○	○	○

Question 3: Learning about terminology management tools

Using the scale provided, please indicate the extent to which you agree or disagree with each of the following statements.

Terminology management tools...	Strongly Disagree		Neutral		Strongly Agree
1. I have seen how other translators use them.	○	○	○	○	○
2. Many freelance translators use them.	○	○	○	○	○
3. Before deciding whether to use them, I was able to try them out fully.	○	○	○	○	○
4. I was permitted to use them on a trial basis long enough to see what they could do.	○	○	○	○	○
5. I would have no difficulty telling others about what they can do.	○	○	○	○	○
6. I believe I could communicate to others the advantages and disadvantages of using them.	○	○	○	○	○
7. The benefits of using them are apparent to me.	○	○	○	○	○
8. I had ample opportunity to try them out before buying.	○	○	○	○	○
9. I need training in using them more effectively.	○	○	○	○	○
10. I taught myself to use them.	○	○	○	○	○
11. I feel confident enough to teach myself to use new ones.	○	○	○	○	○

Appendix 2. Online survey for CAT tools adopters

Question 4: Impacts of terminology management tools

On the scale provided, please indicate the impact that your use of terminology management tools has had on your work.

Impacts of terminology management tools on...	Large Decrease	Small Decrease	Unchanged	Small Increase	Large Increase
1. My turnover	○	○	○	○	○
2. Size of my customer base	○	○	○	○	○
3. Quality of my translations	○	○	○	○	○
4. My productivity	○	○	○	○	○
5. Volume of work I undertake	○	○	○	○	○
6. Number of clients I have	○	○	○	○	○
7. Volume of work offered to me by clients	○	○	○	○	○
8. Prices I charge for work I undertake	○	○	○	○	○

Question 5: Online tools and online linguistic/reference resources

On the scale provided, please indicate the usefulness of the following online search tools and online linguistic/reference resources.

Usefulness of online terminology tools and resources	Not useful at all	Not very useful	Neutral	Useful	Very useful
1. Monolingual dictionaries & glossaries	○	○	○	○	○
2. Multilingual dictionaries & glossaries	○	○	○	○	○
3. Multilingual terminology databases	○	○	○	○	○
4. Encyclopaedia	○	○	○	○	○
5. Document archives	○	○	○	○	○
6. Corpora	○	○	○	○	○
7. Reference databases (e.g. for subject specialism data)	○	○	○	○	○
8. Search engines	○	○	○	○	○
9. Specialists gateways (e.g. Internet portals with reference resources)	○	○	○	○	○
10. Other:	○	○	○	○	○

For me, the **advantages** of using online tools and linguistic/reference resources are:

For me, the **disadvantages** of using online tools and linguistic/reference resources are:

Some online resources are available only upon payment of a subscription or access charge. Do you use any of these services?

○ Yes   ○ No

Question 6: Sharing your terminology resources

Terminology collections stored in electronic formats can be made available to others. Using the scale below, please indicate your level of involvement in terminology sharing.

How often do you...	Never	Rarely	Occasionally	Frequently	Almost always
1. Share your terminology collections with colleagues	o	o	o	o	o
2. Exchange your terminology collections with colleagues	o	o	o	o	o
3. Buy terminology collections from colleagues	o	o	o	o	o
4. Sell your terminology collections to colleagues	o	o	o	o	o

Part B: Translation memory

This section contains questions about translation memory systems, such as Trados, Déjà Vu, SDLX, and StarTransit.

If you do not use translation memory, please go to Part C.

Question 1: Using translation memory

Which translation memory system(s) do you use?

Using the scale provided, please indicate the extent to which you agree or disagree with each of the following statements about translation memory.

Translation memory...	Strongly Disagree		Neutral		Strongly Agree
1. Enables me to accomplish tasks more quickly.	o	o	o	o	o
2. Improves the quality of work I do.	o	o	o	o	o
3. Makes it easier for me to do my job.	o	o	o	o	o
4. Improve my job performance.	o	o	o	o	o
5. Is overall advantageous in my job.	o	o	o	o	o
6. Enhances my effectiveness in my work.	o	o	o	o	o
7. Gives me greater control over my work.	o	o	o	o	o
8. Increases my productivity.	o	o	o	o	o
9. Is compatible with the type of translation assignments I undertake.	o	o	o	o	o
10. Fits well with the way I like to work.	o	o	o	o	o
11. Is cumbersome to use.	o	o	o	o	o
12. Requires a lot of mental effort.	o	o	o	o	o
13. Is often frustrating.	o	o	o	o	o
14. Does what I want to do easily.	o	o	o	o	o
15. Is easy for me to use.	o	o	o	o	o
16. Was easy for me to learn.	o	o	o	o	o

Appendix 2. Online survey for CAT tools adopters

Question 2: Translation memory and the translation sector

Using the scale provided, please indicate the extent to which you agree or disagree with each of the following statements.

Translation memory...	Strongly Disagree		Neutral		Strongly Agree
1. My clients expect me to use it.	O	O	O	O	O
2. My use of it is voluntary.	O	O	O	O	O
3. Using it improves my image within the translation sector.	O	O	O	O	O
4. Clients prefer to work with translators who use it.	O	O	O	O	O
5. Translators who use it have a high profile in the translation sector.	O	O	O	O	O
6. Having it is a status symbol among translators.	O	O	O	O	O

Question 3: Learning about translation memory

Using the scale provided, please indicate the extent to which you agree or disagree with each of the following statements.

Translation memory...	Strongly Disagree		Neutral		Strongly Agree
1. I have seen how other translators use it.	O	O	O	O	O
2. Many freelance translators use it.	O	O	O	O	O
3. Before deciding whether to use it, I was able to try it out fully.	O	O	O	O	O
4. I was permitted to use it on a trial basis long enough to see what it could do.	O	O	O	O	O
5. I would have no difficulty telling others about what it can do.	O	O	O	O	O
6. I believe I could communicate to others the advantages and disadvantages of using it.	O	O	O	O	O
7. The benefits of using it are apparent to me.	O	O	O	O	O
8. I had ample opportunity to try it out before buying.	O	O	O	O	O
9. I need training in using them more effectively.	O	O	O	O	O
10. I taught myself to use them.	O	O	O	O	O
11. I feel confident enough to teach myself to use another new one.	O	O	O	O	O

Question 4: Impacts of translation memory

On the scale provided, please indicate the impact that your use of translation memory has had on your work.

Impacts of translation memory on...	Large Decrease	Small Decrease	Unchanged	Small Increase	Large Increase
1. My turnover	O	O	O	O	O
2. Size of my customer base	O	O	O	O	O
3. Quality of my translations	O	O	O	O	O
4. My productivity	O	O	O	O	O
5. Volume of work I undertake	O	O	O	O	O
6. Number of clients I have	O	O	O	O	O
7. Volume of work offered to me by clients	O	O	O	O	O
8. Prices I charge for work I undertake	O	O	O	O	O

Question 5: Sharing your translation memories

Translation memories can be shared with others. Using the scale below, please indicate your level of involvement in translation memory sharing.

How often do you...	Never	Rarely	Occasionally	Frequently	Almost always
1. Share your translation memories with colleagues	O	O	O	O	O
2. Exchange your translation memories with colleagues	O	O	O	O	O
3. Buy translation memories from colleagues	O	O	O	O	O
4. Sell your translation memories to colleagues	O	O	O	O	O

Appendix 2. Online survey for CAT tools adopters   197

Part C: Your 'translation toolkit'

Question 1: Your tools

Using the table below, please indicate the software tools you use in your translation work.

Task	Tools
Translation production and editing (e.g. MS Word)	
Terminology searches (e.g. search engines, online glossaries)	
Glossary creation (e.g. MultiTerm, Excel)	
Word count	
File management	
Project management	
Text alignment	
Communicating with clients	
File transfer	
Invoice generation	
Book keeping / accounts	
Other task:	

Question 2: External influences

Using the scale below, please indicate the factors that influence your adoption of software tools into your 'translation toolkit'.

It is my perception that pressure to adopt new software tools comes from...	No pressure at all				Total pressure
1. Direct clients / Translation agencies	O	O	O	O	O
2. Software vendors	O	O	O	O	O
3. Other translators	O	O	O	O	O
4. Translation associations / professional bodies	O	O	O	O	O

Question 3: Websites, marketplaces, and discussion groups

Do you have your **own web site** to promote your translation services?

        ○ Yes   ○ No

If no, please click here to continue.

Please tell us about your web site	Me	Other people (please specify)
My web site was created by	○	
My web site content is updated by	○	
My web site design is updated by	○	

In which year was your website created? [ ]

The benefits I have gained from having my own website are:

[ ]

The problems I have encountered with having my own website are:

[ ]

Do you use **online marketplaces/auctions** to bid for translation assignments?

        ○ Yes   ○ No

If no, please click here to continue.

For me, the advantages of using online marketplaces/auctions are:

[ ]

For me, the disadvantages of using online marketplaces/auctions are:

[ ]

Which electronic **mailing lists** and/or **discussion groups** are you subscribed to?

[ ]

If you do not participate in mailing lists or discussion groups, please click here to go to next section.

For me, the advantages of being involved in electronic mailing lists/discussion groups are:

[ ]

The disadvantages for me of being involved in electronic mailing lists/discussion groups are:

[ ]

Appendix 2. Online survey for CAT tools adopters

Part D: Your profile

Question 1: The translation assignments you undertake

Please provide us with some details about the translation assignments you undertake.

Which language pairs do you translate?

FROM	TO

Which subject areas do you translate?

Which document types do you translate (e.g. manuals, technical reports, contracts, patents)?

Approximately what proportion of your workload is delivered to you by e-mail?

○ 0% ○ 1%-25% ○ 25%-50% ○ 50%-75% ○ 75%-99% ○ 100%

Approximately what proportion of your translation assignments do you submit by e-mail?

○ 0% ○ 1%-25% ○ 25%-50% ○ 50%-75% ○ 75%-99% ○ 100%

Which document formats do you usually work with? Please tick all that apply:

- ☐ Rich Text Format (RTF)
- ☐ Word documents (DOC)
- ☐ Plain text (TXT)
- ☐ Wordperfect (WPD)
- ☐ PowerPoint presentations
- ☐ Adobe PDF
- ☐ FrameMaker
- ☐ QuarkXPress
- ☐ PageMaker
- ☐ SGML / XML
- ☐ Web page files (HTML/ASP)
- ☐ Resource files (RC)
- ☐ Source code files (C/C++/Java/VB)
- ☐ Excel spreadsheets (XLS)
- ☐ Other:

What proportion of your work do you undertake for:

    direct clients [   ] %
    translation agencies [   ] %

What is the approximate average size of the translation assignments you undertake?

○ under 1000 words   ○ 1000-5000 words   ○ 5000-10000 words   ○ over 10000 words

Approximately, how many words do you translate per week?

○ under 1000 words   ○ 1000-5000 words   ○ 5000-10000 words   ○ over 10000 words

Question 2: Being a freelancer

Relative to the rest of the freelance translation sector in the UK, how do you rate your performance in the following areas?

	Very weak	Weak	Same level	Strong	Very Strong
1. Long term profitability	○	○	○	○	○
2. Amount of translation work undertaken	○	○	○	○	○
3. Financial resources (liquidity and investment capacity)	○	○	○	○	○
4. Client base	○	○	○	○	○
5. Professional image and client loyalty	○	○	○	○	○

Question 3: Learning to use new software tools

Using the scale below, please indicate your preferences for learning to use new software tools to support you in your translation work.

I would like to learn to use software tools through...	Strongly Disagree		Neutral		Strongly Agree
1. teaching myself	○	○	○	○	○
2. taught courses	○	○	○	○	○
3. workshops for translators	○	○	○	○	○
4. e-learning	○	○	○	○	○
5. training provided by software vendors	○	○	○	○	○

Appendix 2. Online survey for CAT tools adopters   201

Additional comments

Please use this space for any comments you wish to make related to this study. Also, could you include any comments on translation tools and resources that you find have not been fully covered in the study.

Thank you for your help.
To receive your £10 book token, please provide your name and address.

Submit

Appendix 3. Online survey for CAT tools non-adopters

Translation Tools in the 21st century

Part A: Terminology management tools

This section contains questions about terminology management tools, i.e. software packages used for creating and managing your own terminology collections. Examples include MultiTerm, Lingo, TermWatch, and StarTerm.

Question 1: Terminology management tools

Using the scale provided, please indicate the extent to which you agree or disagree with each of the following statements about terminology management tools.

Terminology management tools...	Strongly Disagree		Neutral		Strongly Agree
1. Would enable me to accomplish tasks more quickly.	O	O	O	O	O
2. Would improve the quality of work I do.	O	O	O	O	O
3. Would make it easier for me to do my job.	O	O	O	O	O
4. Would improve my job performance.	O	O	O	O	O
5. Would overall be advantageous in my job.	O	O	O	O	O
6. Would enhance my effectiveness in my work.	O	O	O	O	O
7. Would give me greater control over my work.	O	O	O	O	O
8. Would increase my productivity.	O	O	O	O	O
9. Would be compatible with the type of translation assignments I undertake.	O	O	O	O	O
10. Would fit well with the way I like to work.	O	O	O	O	O
11. Would be cumbersome to use.	O	O	O	O	O
12. Would require a lot of mental effort.	O	O	O	O	O
13. Would often be frustrating.	O	O	O	O	O
14. Would do what I want to do easily.	O	O	O	O	O
15. Would be easy for me to use.	O	O	O	O	O
16. Would be easy for me to learn.	O	O	O	O	O

Question 2: Terminology management tools and the translation sector

Using the scale provided, please indicate the extent to which you agree or disagree with each of the following statements.

Terminology management tools...	Strongly Disagree		Neutral		Strongly Agree
1. My clients expect me to use them.	○	○	○	○	○
2. Using them would improve my image within the translation sector.	○	○	○	○	○
3. Clients prefer to work with translators who use them.	○	○	○	○	○
4. Translators who use them have a high profile in the translation sector.	○	○	○	○	○
5. Having them is a status symbol among translators.	○	○	○	○	○

Question 3: Learning about terminology management tools

Using the scale provided, please indicate the extent to which you agree or disagree with each of the following statements.

Terminology management tools...	Strongly Disagree		Neutral		Strongly Agree
1. I have seen how other translators use them.	○	○	○	○	○
2. Many freelance translators use them.	○	○	○	○	○
3. Before deciding whether to use them, I would be able to try them out fully.	○	○	○	○	○
4. I would be able to use them on a trial basis long enough to see what they could do.	○	○	○	○	○
5. I would have no difficulty telling others about what they can do.	○	○	○	○	○
6. I believe I could communicate to others the advantages and disadvantages of using them.	○	○	○	○	○
7. The benefits of using them are apparent to me.	○	○	○	○	○
8. I would have ample opportunity to try them out before deciding to adopt.	○	○	○	○	○
9. I know where I can go to try out several of them.	○	○	○	○	○
10. I would feel confident enough to teach myself to use them.	○	○	○	○	○

Appendix 3. Online survey for CAT tools non-adopters

Question 4: Impacts of terminology management tools

On the scale provided, please indicate the impact that you believe using terminology management tools would have on your work.

Impacts of terminology management tools on...	Large Decrease	Small Decrease	Unchanged	Small Increase	Large Increase
1. My turnover	○	○	○	○	○
2. Size of my customer base	○	○	○	○	○
3. Quality of my translations	○	○	○	○	○
4. My productivity	○	○	○	○	○
5. Volume of work I undertake	○	○	○	○	○
6. Number of clients I have	○	○	○	○	○
7. Volume of work offered to me by clients	○	○	○	○	○
8. Prices I charge for work I undertake	○	○	○	○	○

Question 5: Online tools and online linguistic/reference resources

On the scale provided, please indicate the usefulness of the following online search tools and online linguistic/reference resources.

Usefulness of online terminology tools and resources	Not useful at all	Not very useful	Neutral	Useful	Very useful
1. Monolingual dictionaries & glossaries	○	○	○	○	○
2. Multilingual dictionaries & glossaries	○	○	○	○	○
3. Multilingual terminology databases	○	○	○	○	○
4. Encyclopaedia	○	○	○	○	○
5. Document archives	○	○	○	○	○
6. Corpora	○	○	○	○	○
7. Reference databases (e.g. for subject specialism data)	○	○	○	○	○
8. Search engines	○	○	○	○	○
9. Specialists gateways (e.g. Internet portals with reference resources)	○	○	○	○	○
10. Other:	○	○	○	○	○

For me, the **advantages** of using online tools and linguistic/reference resources are:

For me, the **disadvantages** of using online tools and linguistic/reference resources are:

Some online resources are available only upon payment of a subscription or access charge. Do you use any of these services?

○ Yes ○ No

Question 6: Sharing your terminology resources

Terminology collections stored in electronic formats can be made available to others. Using the scale below, please indicate your level of involvement in terminology sharing.

How often do you...	Never	Rarely	Occasionally	Frequently	Almost always
1. Share your terminology collections with colleagues	O	O	O	O	O
2. Exchange your terminology collections with colleagues	O	O	O	O	O
3. Buy terminology collections from colleagues	O	O	O	O	O
4. Sell your terminology collections to colleagues	O	O	O	O	O

Part B: Translation memory

This section contains questions about translation memory tools, such as Trados, Déjà Vu, SDLX, and StarTransit.

Question 1: Using translation memory

Using the scale provided, please indicate the extent to which you agree or disagree with each of the following statements about translation memory.

Translation memory...	Strongly Disagree		Neutral		Strongly Agree
1. Would enable me to accomplish tasks more quickly.	O	O	O	O	O
2. Would improve the quality of work I do.	O	O	O	O	O
3. Would make it easier for me to do my job.	O	O	O	O	O
4. Would improve my job performance.	O	O	O	O	O
5. Would be overall advantageous in my job.	O	O	O	O	O
6. Would enhance my effectiveness in my work.	O	O	O	O	O
7. Would give me greater control over my work.	O	O	O	O	O
8. Would increase my productivity.	O	O	O	O	O
9. Would be compatible with the type of translation assignments I undertake.	O	O	O	O	O
10. Would fit well with the way I like to work.	O	O	O	O	O
11. Would be cumbersome to use.	O	O	O	O	O
12. Would require a lot of mental effort.	O	O	O	O	O
13. Would often be frustrating.	O	O	O	O	O
14. Would do what I want to do easily.	O	O	O	O	O
15. Would be easy for me to use.	O	O	O	O	O
16. Would be easy for me to learn.	O	O	O	O	O

Question 2: Translation memory and the translation sector

Using the scale provided, please indicate the extent to which you agree or disagree with each of the following statements.

Translation memory...	Strongly Disagree		Neutral		Strongly Agree
1. My clients expect me to use it.	O	O	O	O	O
2. Using it would improve my image within the translation sector.	O	O	O	O	O
3. Clients prefer to work with translators who use it.	O	O	O	O	O
4. Translators who use it have a high profile in the translation sector.	O	O	O	O	O
5. Having it is a status symbol among translators.	O	O	O	O	O

Question 3: Learning about translation memory

Using the scale provided, please indicate the extent to which you agree or disagree with each of the following statements.

Translation memory...	Strongly Disagree		Neutral		Strongly Agree
1. I have seen how other translators use it.	O	O	O	O	O
2. Many freelance translators use it.	O	O	O	O	O
3. Before deciding whether to use it, I would be able to try it out fully.	O	O	O	O	O
4. I would be permitted to use it on a trial basis long enough to see what it could do.	O	O	O	O	O
5. I would have no difficulty telling others about what it can do.	O	O	O	O	O
6. I believe I could communicate to others the advantages and disadvantages of using it.	O	O	O	O	O
7. The benefits of using it are apparent to me.	O	O	O	O	O
8. I would have ample opportunity to try it out before deciding to adopt.	O	O	O	O	O
9. I know where I can go to try out various translation memory systems.	O	O	O	O	O
10. I would feel confident enough to teach myself to use them.	O	O	O	O	O

Question 4: Impacts of translation memory

On the scale provided, please indicate the impact that you believe using translation memory would have on your work.

Impacts of translation memory on...	Large Decrease	Small Decrease	Unchanged	Small Increase	Large Increase
1. My turnover	O	O	O	O	O
2. Size of my customer base	O	O	O	O	O
3. Quality of my translations	O	O	O	O	O
4. My productivity	O	O	O	O	O
5. Volume of work I undertake	O	O	O	O	O
6. Number of clients I have	O	O	O	O	O
7. Volume of work offered to me by clients	O	O	O	O	O
8. Prices I charge for work I undertake	O	O	O	O	O

Part C: Your 'translation toolkit'

Question 1: Your tools

Using the table below, please indicate the software tools you use in your translation work.

Task	Tools
Translation production and editing (e.g. MS Word)	
Terminology searches (e.g. search engines, online glossaries)	
Glossary creation (e.g. MultiTerm, Excel)	
Word count	
File management	
Project management	
Text alignment	
Communicating with clients	
File transfer	
Invoice generation	
Book keeping / accounts	
Other task:	

Question 2: External influences

Using the scale below, please indicate the factors that influence your adoption of software tools into your 'translation toolkit'.

It is my perception that pressure to adopt new software tools comes from...	No pressure at all				Total pressure
1. Direct clients / Translation agencies	O	O	O	O	O
2. Software vendors	O	O	O	O	O
3. Other translators	O	O	O	O	O
4. Translation associations / professional bodies	O	O	O	O	O

Appendix 3. Online survey for CAT tools non-adopters 209

Question 3. Websites, marketplaces, and discussion groups

Do you have your **own web site** to promote your translation services?

                ○ Yes   ○ No

If no, please click here to continue.

Please tell us about your web site	Me	Other people (please specify)
My web site was created by	○	
My web site content is updated by	○	
My web site design is updated by	○	

In which year was your website created? [ ]

The benefits I have gained from having my own website are:

[ ]

The problems I have encountered with having my own website are:

[ ]

Do you use **online marketplaces/auctions** to bid for translation assignments?

                ○ Yes   ○ No

If no, please click here to continue.

For me, the advantages of using online marketplaces/auctions are:

[ ]

For me, the disadvantages of using online marketplaces/auctions are:

[ ]

Which electronic **mailing lists** and/or **discussion groups** are you subscribed to?

[ ]

If you do not participate in mailing lists or discussion groups, please click here to go to next section.

For me, the advantages of being involved in electronic mailing lists/discussion groups are:

[ ]

The disadvantages for me of being involved in electronic mailing lists/discussion groups are:

[ ]

Part D: Your profile

Question 1: The translation assignments you undertake

Please provide us with some details about the translation assignments you undertake.

Which language pairs do you translate?

FROM	TO

Which subject areas do you translate?

Which document types do you translate (e.g. manuals, technical reports, contracts, patents)?

Approximately what proportion of your workload is delivered to you by e-mail?

○ 0%   ○ 1%-25%   ○ 25%-50%   ○ 50%-75%   ○ 75%-99%   ○ 100%

Approximately what proportion of your translation assignments do you submit by e-mail?

○ 0%   ○ 1%-25%   ○ 25%-50%   ○ 50%-75%   ○ 75%-99%   ○ 100%

Which document formats do you usually work with? Please tick all that apply:

- ☐ Rich Text Format (RTF)
- ☐ Word documents (DOC)
- ☐ Plain text (TXT)
- ☐ Wordperfect (WPD)
- ☐ PowerPoint presentations
- ☐ Adobe PDF
- ☐ FrameMaker
- ☐ QuarkXPress
- ☐ PageMaker
- ☐ SGML / XML
- ☐ Web page files (HTML/ASP)
- ☐ Resource files (RC)
- ☐ Source code files (C/C++/Java/VB)
- ☐ Excel spreadsheets (XLS)
- ☐ Other:

Appendix 3. Online survey for CAT tools non-adopters

What proportion of your work do you undertake for:

    direct clients [ ] %
    translation agencies [ ] %

What is the approximate average size of the translation assignments you undertake?

○ under 1000 words   ○ 1000-5000 words   ○ 5000-10000 words   ○ over 10000 words

Approximately, how many words do you translate per week?

○ under 1000 words   ○ 1000-5000 words   ○ 5000-10000 words   ○ over 10000 words

Question 2: Being a freelancer

Relative to the rest of the freelance translation sector in the UK, how do you rate your performance in the following areas?

	Very weak	Weak	Same level	Strong	Very Strong
1. Long term profitability	○	○	○	○	○
2. Amount of translation work undertaken	○	○	○	○	○
3. Financial resources (liquidity and investment capacity)	○	○	○	○	○
4. Client base	○	○	○	○	○
5. Professional image and client loyalty	○	○	○	○	○

Question 3: Learning to use new software tools

Using the scale below, please indicate your preferences for learning to use new software tools to support you in your translation work.

I would like to learn to use software toolsta through...	Strongly Disagree		Neutral		Strongly Agree
1. teaching myself	○	○	○	○	○
2. taught courses	○	○	○	○	○
3. workshops for translators	○	○	○	○	○
4. e-learning	○	○	○	○	○
5. training provided by software vendors	○	○	○	○	○

Additional comments

Please use this space for any comments you wish to make related to this study. Also, could you include any comments on translation tools and resources that you find have not been fully covered in the study.

Thank you for your help.
To receive your £10 book token, please provide your name and address.

Submit

Appendix 4. Addressing non-response bias: Mann-Whitney test between early and late respondents

1.1. Appendix 4. Addressing non-response bias: Mann-Whitney test between early and late respondents

Result of Mann-Withney test between early and late respondents

Variables tested	Mann-Whitney U	2-tailed significance	Are they significant at 95% level?
Age range	319.0	.045	Significant
Gender	435.0	.792	Not significant
Educational level	400.5	.740	Not significant
Combinations of translation quals	390.0	.305	Not significant
Length of translation experience	422.5	.827	Not significant
Software apps FAM+KNOW: Word processing	375.5	.950	Not significant
Software apps USAGE: Word processing	435.0	.317	Not significant
Software apps FAM+KNOW: Spreadsheet	336.5	.441	Not significant
Software apps USAGE: Spreadsheet	381.5	.615	Not significant
Software apps FAM+KNOW: Database	270.5	.711	Not significant
Software apps USAGE: Database	287.0	.767	Not significant
Software apps FAM+KNOW: Accounting	254.0	.775	Not significant
Software apps USAGE: Accounting	263.5	.530	Not significant
Software apps FAM+KNOW: Desktop publishing	264.5	.318	Not significant
Software apps USAGE: Desktop publishing	314.0	.728	Not significant
Software apps FAM+KNOW: Web publishing	269.5	.860	Not significant
Software apps USAGE: Web publishing	288.0	1.000	Not significant
Software apps FAM+KNOW: Graphics	268.5	.863	Not significant
Software apps USAGE: Graphics	267.5	.523	Not significant
Software apps FAM+KNOW: Info retrieval+OCR	265.5	.791	Not significant
Software apps USAGE: Info retrieval + OCR	270.5	.430	Not significant
Software apps FAM+KNOW: Groupware	209.5	.064	Not significant
Software apps USAGE: Groupware	264.5	.328	Not significant
Software apps FAM+KNOW: Project mgment	255.0	.440	Not significant
Software apps USAGE: Project mgment	276.0	1.000	Not significant
Software apps FAM+KNOW: Terminology	232.0	.090	Not significant
Software apps USAGE: Terminology	275.0	.338	Not significant
Software apps FAM+KNOW: MT	246.0	.580	Not significant
Software apps USAGE: MT	253.0	.338	Not significant
Software apps FAM+KNOW: CAT	277.0	.476	Not significant
Software apps USAGE: CAT	270.5	.184	Not significant

Software apps FAM+KNOW: Localisation	231.5	.322	Not significant
Software apps USAGE: Localisation	253.0	.338	Not significant
IT opinion: prev. exp. necessary	430.5	.742	Not significant
IT opinion: computerisation = higher quality services	420.0	.623	Not significant
IT opinion: computerisation = time saving benefits	395.5	.371	Not significant
IT opinion: + benefits IF integrated apps	385.5	.313	Not significant
IT opinion: computerisation = + effectiveness as translator	357.5	.132	Not significant
IT opinion: computerisation = + revenue	347.0	.103	Not significant
IT opinion: computerisation = + comms with customers	361.5	.154	Not significant
IT opinion: apps failed to meet requirements	353.5	.191	Not significant
IT opinion: computerisation = many problems	407.5	.504	Not significant
IT opinion: use of apps = failure so far	421.0	.808	Not significant
IT opinion: computerisation = - benefits than expected	354.0	.191	Not significant

Appendix 5. Qualitative analysis form

QUALITATIVE ANALYSIS DOCUMENTATION FORM

1. Research Issue being explored: *Training, Continuous Professional Development (CPD) and successful adoption of TTs* Analyst: XG Version: 1/1

2. Aims of analysis: Find relationships between translators' training in TTs, their "Continuous Professional Development" and the successful adoption of TTs (obviously for those translators who have already adopted) + check if comparison/relation can be made/established between AD and NA (finding group of "likely NA for successful adoption"??).

3. Description of procedures:

DATA SETS IN USE	PROCEDURAL STEPS (Numbered, what was done and how it was done)	DECISION RULES Followed during analysis operations	ANALYSIS OPERATIONS (enter codes)				CONCLUSIONS DRAWN	RESEARCHER COMMENTS
			Readying data for analysis	Drawing conclusions	Confirming conclusions			
adopters19_SPSS.sav	*1) Asking the prediction question:* How does Training and Continuous Professional Development (CPD) relate to the successful adoption of TTs?			-	-		-	
adopters19_SPSS.sav → "analysis matrices.xls" 1AD-Partially ordered matrix	*2) Selecting the predictors:* For successful adoption: - TMTs/TMs/TTs Impact scores - Individual Impact variables (A_IMPACT1-8 and B_IMPACT1-8). For training in TTs and CPD: - Self-confidence variables (A_SELFCO1-3 and B_SELFCO1-3) - Variables measuring Q3 (Learning to use new software tools) in Part D of web survey (labelled as: D32_self, D33_taught, D34_workshops, D35_e-learning, and D35_vendors).	1st Data collected from each case into *Partially ordered matrix* for data formatting/ standardising/reducing. 2nd Re-ordering of variables to place predictors at the beginning.	MAT SUB COMP	-	-		-	

Appendix 5. Qualitative analysis form

"analysis matrices.xls" 1AD Partially ordered matrix	3) *Scaling the outcome and the predictors*: The dependent variable (adoption decision) was already scaled/split into the dichotomy *adopted TTs* or *not adopted TTs*. The predictors measuring TTs adoption success were used to determine the degree of the success of the adoption. 2 cases showed a negative impact of TTs, while the remaining 17 showed a moderate (15) or high (2 cases) positive impact of TTs. With regard to the predictors looking at training and CPD, responses in the shape of 1–5 variations of a Likert scale were scaled to 3 categories (disagree/neutral/agree). Also, 2 new variables were created as a result of subsuming impacts' variables affecting TMTs (*TMTs Impact score*), TMs (*TMs Impact score*), and another new variable (*TTs Impact score*) subsumed the values of the previous two new variables.	Scaling decisions: 1st AD/NA 2nd Scaling 5-points Likert scales for Impacts' variables to 5 "labelled" statements: - *variable with high negative impact upon adoption* (values < 2) - *variable with moderate negative impact upon adoption* (values between 2 and 3) - *variable with no impact upon adoption* (value 3) - *moderate positive impact upon adoption* (values between 3 and 4) - *variables with high positive impact upon adoption* (values above 4). 3rd To subsume *TMTs impacts score* and *TMs impacts score*, "combined variables scores" (arithmetic mean values) of the values of the adoption factors corresponding to each technology were computed. Similarly, to work out the *TTs impacts score*, "combined variables scores" (arithmetic mean values) of the previous subsumed indexes was computed. 4th Scaling 5-points Likert scales for training and CPD variables to 3 categories: - *disagreement* (values between 1 and 2.3) - *neutrality* (values between 2.3 and 3.6) - *agreement* (values between 3.6 and 5).	CLAS SCAL SUMM COMP	Use of spreadsheet for matrix building would allow better manipulation of data than an SPSS file.

→ 2AD Case-ordered descriptive matrix	4) *Building the matrix and entering data*: A matrix with predictors was built and the rest of the variables' information was kept at the end of each case's row. Colour codes were used for each predictor category's value (see row above). Once data was entered, a case by case review of the data sets was performed. See comments' column for details.	Conversion of partially ordered meta-matrix into a *Case-ordered descriptive meta-matrix*. Starting to notice patterns. Initially, only adopters would be included in the matrix building process. Later exploration of relationships/comparisons with NA's perspectives may entail building another matrix for NA. Keep all data from each case, but key analysis variables were located at the beginning of each case, after its id information. Use of colour codes to ease patterns identification. Cases ad02 and ad13 were not using TMTs, thus they did not report any Impact of TMTs. As a consequence, they should be ignored for overall conclusions.	MAT RANK
→ 3AD5 Case-ordered predictor matrix	5) *Drawing first conclusions*: Search of useful information looking down the columns.	Conversion of *Case-ordered descriptive meta-matrix* into *Case-ordered predictor-outcome matrix* to see whether antecedent variables (training and CPD predictors) account for criterion variable (adoption success).	RANK PAT COUNT CONT

Appendix 5. Qualitative analysis form

→ 3AD5 Case-ordered predictor matrix	6) *Testing the prediction:* How does Training and Continuous Professional Development (CPD) relate to the successful adoption of TTs?	1) Computing "combined variables scores" used to order cases and to help finding relationships. 2) Key RQ related to prediction question: ??	COMP	RANK PAT COUNT CONT REL	TBD	Among adopters of TTs, there was an overall successful adoption of technologies, with diverse degrees of success. The majority of the translators showed overall moderate levels of success, and just two cases showed a highly successful adoption of TTs. The two translators who most successfully adopted TTs* did not feel they needed more training in using them more effectively, although one taught herself to use them and was also confident enough to teach herself to use new ones, and the other (which was the most successful case of TTs adoption) represented a completely opposite case: not self-taught, nor confident to teach herself to use new tools. Translators who had a moderate success in their adoption of TTs, had broadly taught themselves to use them; however, their level of confidence to learn to use new ones was lower than in the previous group, which could be related to the fact that they felt they needed training to use them more effectively. Preferences among successful adopters of TTs with regard to learning to use new TTs ranked workshops specially designed for translators and taught courses as the preferred learning ways.	Conclusions used for abstract of [09_IS]

Translators who had not adopted TTs mainly used electronic mailing lists/discussion groups for the purpose of collaborating with their colleagues, or discuss terminology problems with them. On the other hand, for those translators who had adopted TTs, such perceived advantages came only after using electronic mailing lists/discussion groups for keeping up to date in the sector or being aware of the latest developments.

Translators with a moderate success in their adoption of TTs seemed to be using online collaboration resources for translators (such as translation marketplaces or translation mailing lists) more extensively than those who had a rather higher success in the adoption of TTs, and seemed to be more eager to keep abreast of new technological developments in TTs and the translation sector.

The last two statements could imply that the translators make more use of electronic resources to catch up with new developments as they adopt TTs, and once they reach a considerable level of success in their adoption of TTs, their interest seems to decrease again.

Appendix 5. Qualitative analysis form

CODE LIST FOR ANALYSIS OPERATIONS (Adapted from Miles and Huberman 1994, p. 285)

Readying data for analysis		Drawing conclusions		Confirming conclusions	
MAT	filling in matrices	PLAUS	seeing "plausibility" only	REPR	checking for representativeness
CLAS	classifying, categorising	PAT	noting patterns/themes	RES-EFF	checking for researcher effects
RANK	ranking/weighting data	CLUS	clustering	TRI	triangulation
				TRI-DATA	from different data sources
				TRI-METH	from different methods
				TRI-CONC	conceptually (different theories)
				TRI-RES	from different researchers
SUMM	summarising phrases, generating key words	MET	making metaphors	WT	weighting the evidence
SUB	subsuming data under higher level variable	COUNT	counting/frequencies	OUT	use of outliers, exceptions
SCAL	scaling, summing indices	CEN	establishing central tendencies	EXTR-SIT	extreme situation verification
COMP	computing	CONT	making contrasts/comparisons	EXTR-BIAS	extreme bias verification
SPLT	splitting one variable into two	FAC	establishing factors	SURP	following up surprises
PAR	partitioning	REL	establishing relationships between variables/sets of variables	EMP	empirical evidence from elsewhere
AGG	aggregating	INTV	establishing intervening/linking conditions	NONEG	absence of negative evidence
		LOG	logical chain of evidence	IF-THEN	testing if-then relationships
		COH	making conceptual/theoretical coherence	FALSE-REL	checking false relation due to third variable
				REPL	replication
				RIV	test of rival explanation
				FEEDB	corroboration from informant feedback

TTs, Translation Technologies; TMT(s), Terminology Management Tool(s); TM(s), Translation Memory(-ies)

Appendix 6. Summary of qualitative data analysis

1. **DATA COLLECTION**
   - Online survey design
   - Online survey implementation
   - Data reception

2. **DATA REDUCTION**
   - Data logging and processing (email messages with responses > csv file using perl script)
   - Data import to SPSS and Excel for manipulation
   - Creation of variables for constructs (Subsuming data under higher level variables) [SPSS]

3. **DATA DISPLAYS**
   - Entering data into matrices (partially ordered meta-matrix with all variables, all cases) [Excel]
   - Creation of case-ordered descriptive meta-matrix (All first level descriptive data, form all cases ordered by key variable/s, grouped at beginning of file/matrix)
     - Matrix 1: Factors affecting CAT tool adoption of AD
     - Matrix 2: Factors affecting CAT tool adoption of NA
   - Creation of case-ordered predictor-outcome matrices (Cases arrayed by criterion variable, providing data on main variables – constructs – that may contribute to the outcome)
     - Matrix 1: COMP affecting CAT tool adoption of AD
     - Matrix 2: COMP affecting CAT tool adoption of NA

4. **CONCLUSION DRAWING/VERIFICATION**
   1. Asking the prediction question
      - What factors are associated with the main variable?
   2. Selecting the predictors
      - Perceived predictors, i.e. factors that may have an effect on the main variable.
   3. Scaling the outcome and the predictors
      - Ordering degrees of the outcome variable is a fairly straightforward operation. Cross-case analysts often face this need to "scale" – or otherwise standardise – single-case data; that is, transforming the single-case data from a nominal or categorical variable to a dichotomous or continuous one.
      - This is not an operation to be taken lightly. Rather, it should be a self-conscious procedure with clear decision rules.
   4. Building the matrix and entering data
      - With a clear list of predictors, the construction of the matrix is straightforward.
      - Analyst works case by case, reviewing the case matrices and forming a judgement of the degree to which each of the predictors was in place.
      - Second analyst should verify or disconfirm such judgements.
   5. Drawing first conclusions
      - Look down the columns and seek for useful information (tactics: seeing patterns, making contrasts, comparisons)

6. Testing the prediction
    - What predicts the main variable?
        - TEST1:
            - When cases are ordered by a key variable, and a prediction is made upon the variables this key variable and the rest, patterns should show this relationship. However, should this relationship not be linear, the columns patterns would not show an expected progressive pattern, and thus the prediction could not be confirmed.
            - Part of the reason the prediction would not work out might be limited variation in some of the predictors, where most cases may have a similar rating.
            - Scaled matrices such as these live or die by the variance of their scales. Without variation, the matrix is out of business for other than descriptive purposes. Finding what does not predict is not much help.
        - TEST2:
            - Another test of the prediction can be made by converting the rating (symbols/labels/zero values) to numbers. Computing "combined variables scores" and "group medians" can help to find relationships. If they are not still clear, case ordering should be fine-tuned, and if even then it is not clear, maybe look at some other variable that helps understanding the lack of clarity.
            - So the numbers help. They make it easier to manipulate the data in a case-ordered matrix when you're testing a hypothesis. They also make it easier for you to see what is there. Still, numbers are present along with the rest of the cases' data. Also, the numbers and weights are, thankfully, primitive ones, which keeps you from veering off into fancier data manipulations, which would almost surely be tepid or misleading.

Index

A

activity, 11, 12, 15, 16, 17, 34, 45, 47, 48, 49, 67, 68, 70, 71, 72, 74, 75, 76, 78, 80, 81, 82, 83, 106, 114, 118, 120, 121, 122, 126, 129, 130, 156, 162, 164, 167
  activities, 12, 13, 14, 17, 18, 19, 23, 31, 32, 37, 38, 43, 45, 69, 70, 71, 72, 84, 85, 86, 88, 107, 110, 111, 115, 118, 120, 121, 122, 125, 126, 127, 130, 158, 163, 164, 167
adopter, 93, 101, 107, 109, 110, 148, 149, 153, 167
adoption, 1, 2, 22, 31, 32, 33, 34, 45, 50, 51, 53, 54, 55, 56, 57, 58, 60, 64, 65, 66, 67, 68, 69, 70, 71, 85, 86, 87, 88, 90, 91, 92, 93, 94, 95, 96, 99, 102, 103, 106, 107, 109, 110, 111, 113, 114, 116, 118, 125, 126, 127, 128, 129, 130, 132, 133, 134, 135, 136, 147, 148, 150, 151, 152, 154, 167

B

business management, 5, 45, 50, 65, 66, 68, 70, 97, 118, 125, 127, 129, 167

C

Computer-Assisted Translation, 25, 28, 167
  CAT tools, 1, 2, 22, 23, 24, 26, 27, 31, 32, 33, 34, 54, 57, 64, 65, 67, 68, 70, 75, 77, 88, 93, 95, 98, 102, 103, 105, 107, 109, 110, 111, 113, 114, 115, 116, 118, 121, 125, 126, 127, 128, 129, 130, 132, 133, 135, 136, 139, 140, 141, 142, 143, 144, 145, 147, 148, 149, 150, 151, 152, 153, 154, 163, 167, 168
  CAT tool perceptions, 111, 144, 167
communication, 1, 3, 5, 6, 7, 8, 9, 10, 11, 12, 13, 16, 17, 18, 21, 31, 35, 40, 42, 45, 49, 71, 77, 78, 79, 88, 94, 98, 100, 118, 122, 127, 129, 138, 139, 151, 158, 162, 163, 167
communications, 9, 12, 13, 14, 23, 35, 45, 48, 49, 65, 66, 70, 75, 79, 100, 143, 163, 167
community of practice, 1, 2, 17, 40, 167
compatibility, 30, 103, 106, 113, 147, 148, 149, 150, 151, 167
competences, 6, 16, 17, 35, 37, 39, 42, 43, 44, 49, 167
conclusion drawing, 111, 113, 167
content curation, 75, 80, 81, 82, 161, 167

D

data analysis, 90, 106, 107, 111, 112, 116, 167
data display, 111, 114, 167
data management, 83, 84, 85, 167
data reduction, 99, 111, 114, 147, 167
document production, 32, 66, 70, 85, 97, 118, 120, 122, 127, 129, 130, 167

E

ease of use, 101, 103, 113, 147, 148, 167

I

ICT sophistication, 58, 66, 68, 94, 99, 154, 167
Information and Communication Technology, 50, 167
  Information Technology usage, 96, 167
information competence, 167
information literacy, 2, 3, 5, 35, 36, 38, 39, 40, 42, 43, 65, 69, 156, 162, 164, 167
IL. *See* information literacy
information management, 2, 16, 40, 43, 53, 54, 67, 69, 70, 72, 78, 80, 86, 90, 92, 116, 158, 162, 163, 164, 167
information search and retrieval, 70, 85, 118, 121, 122, 127, 129, 167
information systems, 2, 3, 6, 11, 12, 14, 15, 16, 39, 40, 45, 47, 48, 49, 51, 53, 57, 58, 65, 85, 86, 91, 97, 158, 161, 164, 167
Internet, 1, 23, 26, 31, 34, 35, 36, 39, 65, 73, 74, 94, 96, 97, 101, 106, 110, 162, 167
  Internet usage, 96, 97, 167

IS adoption, 51, 167
IS strategy, 47, 48, 49, 161, 162, 168
IT strategy, 96, 97, 98, 168

K

knowledge, 3, 6, 7, 8, 11, 12, 13, 15, 16, 17, 18, 22, 24, 31, 35, 36, 37, 38, 39, 40, 43, 46, 47, 56, 57, 68, 69, 74, 75, 79, 80, 90, 94, 96, 97, 110, 116, 117, 120, 121, 127, 132, 158, 159, 161, 162, 163, 168

L

lifelong learning, 35, 42, 44, 65, 69, 156, 158, 159, 168

M

machine translation, 24, 25, 27, 32, 67, 75, 77, 121, 122, 127, 128, 168
marketing and work procurement, 70, 118, 125, 127, 129, 168
multilingual communication, 10, 11, 21, 88, 168
Multilingual Information Management System, 6, 72, 156, 158, 160, 162, 164, 168
 MIMS. *See* Multilingual Information Management System
multilingual information profesional, 168

N

non-adopter, 93, 101, 148, 149, 168

P

Personal Learning Environment, 168
 PLE. *See* Personal Learning Environment
professional development, 44, 64, 69, 75, 168

Q

qualitative data analysis, 111, 168
questionnaire, 91, 92, 93, 94, 95, 96, 97, 98, 99, 100, 101, 102, 103, 106, 108, 110, 168

R

relative advantage, 53, 103, 113, 147, 148, 151, 168
result demostrability, 168

S

strategy, 12, 32, 45, 47, 48, 49, 65, 69, 75, 90, 91, 94, 98, 101, 113, 163, 164, 168
survey, 34, 92, 93, 94, 95, 97, 99, 100, 101, 102, 105, 106, 107, 108, 116, 127, 168

T

terminology management system, 24, 168
terminology management tool, 24, 168
translation creation, 67, 70, 85, 118, 121, 125, 126, 127, 129, 168
Translation Memory, 31, 33, 34, 168
translation service provider, 168
translation skills, 16, 168
translation technology, 29, 48, 77, 168
translation toolkit, 102, 168
translation tools, 23, 24, 26, 27, 28, 29, 30, 31, 32, 34, 48, 67, 71, 91, 168
translator, 1, 8, 10, 17, 18, 19, 23, 24, 25, 26, 27, 28, 29, 30, 31, 32, 42, 54, 57, 64, 65, 68, 70, 71, 72, 73, 75, 81, 86, 88, 96, 106, 109, 110, 113, 120, 122, 126, 127, 129, 130, 132, 138, 142, 144, 147, 148, 152, 153, 168
translator profile, 96, 168
translator's workstation, 23, 27, 28, 32, 68, 70, 72, 126, 168
trialability, 103, 113, 147, 148, 169

V

visibility, 48, 83, 103, 113, 147, 148, 169
voluntariness, 103, 113, 147, 148, 151, 169

W

web resources, 43, 169
word processing, 32, 66, 73, 128, 130, 169

Made in the USA
Monee, IL
03 May 2026

49438378R00134